Communications in Asteroseismology

Volume 162
2011

Austrian Academy
of Sciences Press

Vienna 2011 **ÖAW**

Communications in Asteroseismology

Editor-in-Chief: **Michel Breger**, michel.breger@univie.ac.at
Editorial Assistant: **Isolde Müller**, isolde.mueller@univie.ac.at
Layout & Production Manager: **Isolde Müller**, isolde.mueller@univie.ac.at

CoAst Editorial and Production Office
Türkenschanzstraße 17, A - 1180 Wien, Austria
http://www.oeaw.ac.at/CoAst/
Comm.Astro@univie.ac.at

Cover Illustration

Line-profile variations affecting the Si III triplet of the hybrid pulsator γ Pegasi.
(Illustration kindly provided by C. P. Pandey. For more information see the
paper by C. P. Pandey et al., page 21ff.)

British Library Cataloguing in Publication data.
A Catalogue record for this book is available from the British Library.

Austrian Academy of Sciences Press
A-1011 Wien, Postfach 471, Postgasse 7/4
Tel. +43-1-515 81/DW 3402-3406, +43-1-512 9050
Fax +43-1-515 81/DW 3400
http://verlag.oeaw.ac.at, e-mail: verlag@oeaw.ac.at

Contents

Scientific Papers

Scientific
Papers

Comm. in Asteroseismology
Volume 162, November 2010
© Austrian Academy of Sciences

Analyses of Crawford's $uvby\beta$ calibrations using the pulsational variations of FG Vir

P. Haas[1]

[1] Institut für Astronomie, Türkenschanzstrasse 17, A-1180 Vienna, Austria

Abstract

Crawford's $uvby\beta$ calibration method is examined for A-type stars by comparing it with the pulsational variations of the observable m_1, c_1 and M_V for the δ Scuti star FG Vir. The fit between the calibration values of m_1 and M_V and the respective measurements for FG Vir are tested as a function of temperature based on 3068 4-colour values taken at the Observatorio de Sierra Nevada in Spain during the years 2002 and 2003. Testing is performed by means of linear regression. The fit between the measured index m_1 of FG Vir and the m_1 index of the Hyades is nearly perfect. A fit between the calibration value M_V and the measured values of FG Vir cannot be obtained with Crawford's calibration procedure in a straightforward manner. In order to achieve an optimal fit for M_V two modifications of the calibration procedure are investigated and discussed. (i) the position of the ZAMS given by Crawford is replaced by the position of the ZAMS given by Mermilliod; (ii) the influence of the mass difference on c_1 is taken into account.

Accepted: November 5, 2010

Individual Objects: FG Vir

1. Introduction

About five decades ago, Bengt Strömgren (1963, 1966) developed the *uvby* intermediate-band photometric system, which became a widely used tool for determining fundamental parameters of stars. First calibrations of the system were carried out by Strömgren and Crawford (1966, 1975, 1978, 1979) and by Perry, Olsen and Crawford (1987). So far, all studies were based on a large number of stars or sets of stars in order to improve the calibration of the parameters. Several investigations have been reported in the past on how the Strömgren parameters m_1 und c_1 behave during the pulsational variations of a single variable star. Rodríguez et al. (1991) investigated the behaviour of m_1 for several δ Scuti stars, Rodríguez et al. (1992, 1998, 2003) the behaviour of m_1 and c_1 for more δ Scuti stars, Stankov et al. (2002) the behaviour of m_1 and c_1 for the star SX Phe.

For the current investigation the δ Scuti star FG Vir has been used in order to test Crawford's *uvbyβ* calibrations. Various reasons make this star a suitable candidate, for instance:

a) FG Vir is a pulsating star and during its pulsation the star is varying in T_{eff}, log g, etc., and the star's indices, at each moment, must resemble those of one standard star with the same parameters (T_{eff}, log g, M_V and metallicity index m_1),

b) the data set used is very precise,

c) FG Vir is within the range of validity of Crawford's calibrations.

FG Vir is classified as an A5-type star on the Main Sequence (MS) with $< \delta c_1 > = 0.09$ mag as can be seen in Fig. 4; hence D(log g) is about 0.4 from ZAMS, and log g of FG Vir is 3.95. Last but not least, the star has nearly solar abundances: $< \delta m_1 > = 0.018$ as can be derived from Fig. 1; hence [Me/H] = -0.11 follows using Smalley's (1993) calibrations. According to this, one can assume FG Vir as an excellent candidate to test the ZAMS calibrations of Crawford (1979) or similar ones by Philip & Egret (1980).

Changes of luminosity according to the pulsational phase as well as of log g have to be taken into account for the determination of δc_1 relative to the ZAMS as will be outlined in section 7.. For High Amplitude δ Scuti (HADS) and SX Phe pulsators see for example the loops of T_{eff} and log g in the (c_1, $b - y$) diagrams during the pulsation cycle in Fig. 4 of Rodríguez et al. (1993), and Fig. 6 of Rodríguez et al. (2003) for V2109 Cyg. For the star SX Phe in Rolland et al. (1991) Fig. 8 shows an analogous behaviour.

Another question which has to be answered is that of the influence of metallicity [Me/H]. Let us assume a pulsating star similar to FG Vir as far as its location in the HR diagram is concerned but with a quite different metallicity, e.g. [Me/H] = - 3.0. Can we use this star to test Crawford's calibrations for the ZAMS? The answer is negative, because for this star the predictions for m_1 variations during the pulsation cycle, which imply the dependence on temperature, is very different from the predictions for the ZAMS. The prediction is valid for solar abundances. The predictions by the (m_1, β) grids change with T_{eff}, log g and most important, with [Me/H] as Rodríguez et al. (1991) have shown in Fig. 3, i.e. m_1 of the star has to follow the predictions given by the corresponding (m_1, β) grids. And these predictions are quite different for a ZAMS star with solar abundances. Hence, the δm_1 index must vary during the pulsation cycle in agreement with the corresponding predictions. This does not mean that the metallicity of the pulsating star is varying. This means for this star that its m_1-index behaviour during the pulsation cycle is different from that predicted for a ZAMS star with solar abundances.

Even if we consider a pulsating High Amplitude star on the Main Sequence with solar abundances, we can find variations of δm_1 during the pulsation cycle. During the pulsation the surface gravity of the star is varying as well as its luminosity and the m_1-index behaviour depends on surface gravity too as shown by Rodríguez et al. (1991) (see, e.g., Fig. 2 and Fig. 4 of his paper). For high amplitude pulsations the dependence $m_1 = m_1(log\ g)$ must be taken into account to test the ZAMS calibrations.

To summarize: the metallicity index m_1 varies with effective temperature and to a less degree with surface gravity.

In the current investigation Strömgren photometric data for the δ Scuti star FG Vir were collected during 42 nights in the years 2002 and 2003 at the Observatorio de Sierra Nevada. For more details see section 2.

Out of these 42 nights 37 nights have been selected and were combined to an entity of 3068 records, with each record containing the measurements of u, v, b and y. These data can be regarded as data from different stars, since FG Vir is changing its surface temperature T_{eff}, its radius, its surface gravity g etc. during pulsation. On the contrary, the mass of FG Vir and the abundance of all elements remain constant.

Based on these measurements the present paper is testing the calibrations of the metallicity index m_1, the absolute magnitude M_V and the Balmer discontinuity index c_1 according to the tables and procedures given by Crawford (1979) for A-type stars. The observed values of the measured colours should fit the colours and M_V based on Crawford's calibrations. Linear relations between the physical properties of FG Vir are assumed because of its low amplitude pulsating behaviour.

The present paper demonstrates that a fit between the trends of the measured values of y with those of M_V, determined according to Crawford's calibration procedure, can only be achieved with essential modifications to the calibration procedure. Possible modifications of the calibration procedure are discussed.

2. Observations

The investigations in the present paper are based on measurements obtained with the 0.90 m telescope located at 2900 m above sea level in the South-East of Spain at the Observatorio de Sierra Nevada in Granada. The telescope was equipped with the simultaneous four-channel photometer (*uvby* Strömgren photoelectric photometer). For more details see Breger et al. (2004, 2005). In 2002 129 hours of data during 29 nights, and in 2003 24.9 hours during 8 nights were collected. This data set was selected for the final analyses because of its satisfactory small scatter. In total, this sums up to about 154 hours of observation time for both years.

In this paper it is implicitly assumed that the intrinsic values m_0 and c_0 of FG Vir are the same as the observed m_1 and c_1 ones. It is valid in the case of FG Vir because the reddening $E(b - y) = 0$ (see e.g. the reddening assuming the mean indices published in the catalogue of Rodríguez et al. (2000) for δ Scuti-type stars).

Moreover, the given data represent measurements in the *uvby* filters. I determined the β values directly from the corresponding $(b - y)$ ones assuming the $\beta(b - y)$ relation from Crawford (1979) for standard A-type stars. This is valid when, as in the case of FG Vir (more or less),
a) the reddening is null and
b) the star is on the Main Sequence.

If a star is not exactly on the Main Sequence, then a correction for "luminosity" has to be taken into acccount. See the relation in Crawford (1979) when δm_1 is positive:

$$(b - y)_0 = 2.946 - \beta - 0.1\delta c_1$$

In the case of FG Vir $\delta c_1 = 0.09$; hence a difference of about -0.009 mag occurs to the corresponding β values in the ZAMS which can be neglected with respect to the aims of my investigations, dealing with the testing of the trend of m_1 and M_V. Throughout this paper Crawford's (1979) transformation for the standard stars will be used for determining the corresponding β values for each data record, which will be used for all the following analyses.

3.   Testing the temperature dependence of m_1 and δm_1

Crawford (1979) published calibrations of the Strömgren parameters for A-type stars based on measurements of a large number of different stars. The pulsating star FG Vir changes its luminosity, surface temperature and log g during a pulsation cycle. The mass, however, remains constant and the chemical abundance does not change either. Those are the prerequisites of this paper for checking the applicability of Crawford's (1979) calibration procedure through the change of the Strömgren parameter m_1 of the pulsating star FG Vir. In a first step the colour index $m_1(\beta)$ of FG Vir is compared to the colour index $m_1(\beta)$ of the standard stars (Hyades) as published by Crawford (1979).

The method of testing the agreement of the colour indices is a linear fit of $m_1(\beta)$ from which the slope k_m can be determined by:

$$m_1(\beta) = a_m + k_m\beta \,.$$

For better comparability of the two trends and taking into consideration the range of β for FG Vir, the $m_1(\beta)$ values of the standard stars have also been linearized.

Fig. 1 shows the good agreement between the trends of m_1 for the Hyades and FG Vir caused by the pulsations. Let us denote for each β m_{1m} as the " mea-

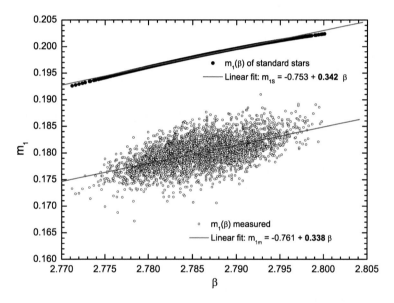

Figure 1: Comparison of the trend of m_1 of the standard stars with the trend of m_1 seen during pulsation of FG Vir. The shift between the two trend lines is caused by a different abundance between the Hyades and FG Vir. This has no effect to the calibration.

sured" values of FG Vir and m_{1S} as the values for the corresponding standard star. Crawford (1975) defined the blanketing parameter $\delta m_1(\beta) = m_{1S}(\beta) - m_{1m}(\beta)$. This parameter can be used for a consistency test. The constance of δm_1 over the pulsation cycle (Fig. 2) supports the conformity of the trend of $m_{1m}(\beta)$ and the trend of $m_{1S}(\beta)$. The conformity of both trends is also shown in Table 1. Data have been divided into 5 contiguous groups each comprising about 610 records.

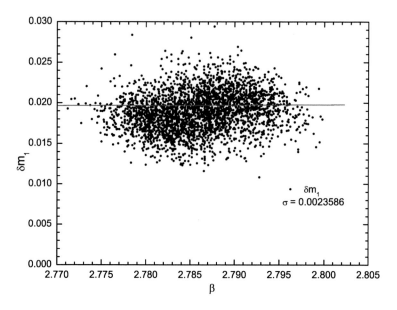

Figure 2: Consistency test based on δm_1

Table 1: Consistency test based on mean values of δm_1

$[\beta]$	$[b-y]$	$[\delta m_1]$	σ
2.779	0.167	0.0182	0.0020
2.783	0.163	0.0179	0.0025
2.786	0.160	0.0180	0.0025
2.789	0.157	0.0186	0.0025
2.793	0.153	0.0181	0.0022

If the size of the measured data sample is sufficiently large, a statement can be made with respect to the behaviour of the photometric index m_1 for δ Scuti stars pulsating with low amplitudes.

Evidently, the conformity of the trend of measured m_1 and the trend of calibrated m_1 depends on the abundances of the metals of the analysed star. Rodríguez et al. (1991) showed that differences in the behaviour of the m_1 variations in some stars with similar surface temperatures and gravities could be caused by different metallicities. Furthermore, Rodríguez (1991) pointed out that some authors assume a varying microturbulence velocity during a pulsation cycle responsible for the changing behaviour of m_1. This affects the strength of the metal-lines. Stankov (2002) showed that for the metal-poor star SX Phe no conformity of the trend of m_1 with the trend of the standard stars can be seen. Rodríguez (1991) explained this phenomenon with the existence of a different trend of m_1 for stars with low metallicities, referring to SX Phe stars explicitly, as a function of the δm_1-index variation and consequently also of the m_1-index variation. Over the pulsation cycle the variation of m_1 increases with decreasing metallicity. Önehag et al. (2009) investigated the calibration of Strömgren *uvby* Hβ photometry for G and F stars and tested the influence of temperature, gravity and metallicity on the m_1 index. The authors reported similar results as those found in this work on A-type stars.

4. Testing the temperature dependence of $c_1(\beta)$

The Balmer discontinuity-index c_1 provides an estimate for the luminosity of a star from the computed difference δc_1 relatively to stars at the ZAMS. For High Amplitude δ Scuti stars studies of the variations of the colour indices c_1 and δc_1 have been performed by Rodríguez et al. (1993 and 2003) and for SX Phe by Rolland et al. (1991). Also a study of the trend of δc_1 can be found in Stankov et al. (2002) for SX Phe. In their paper a value of $\delta V/\delta c_1 = 9.8$ for SX Phe is derived which is in good agreement with the value given in Crawford's (1975) paper for F-type stars.

For FG Vir the procedure, as specified by Stankov (2002), provides a value for $\delta y/\delta c_1 = 10.3$ as can be seen in Fig. 3. Note that by definition y = V. In this plot the band of y is segmented into 9 groups of equidistant width.

The above discussions relate to the behaviour of the Strömgren-index c_1 for the single star FG Vir. The paper of Crawford (1979) describes the calibration of c_1 at constant effective temperature T_{eff} and his results are based on the analysis of different stars with different masses. The dependence of the luminosity on the stellar mass will be shown in the following sections about the M_V calibration. The differences in mass have an influence (even if it is low) on the value of $\log g$ and consequently also on c_1.

5. Relation between δc_1 and δm_1 in A-type stars

Crawford (1979) examined the correlation between the trends of δc_1 and δm_1 for A-type stars and found a small relationship between these two indices: "... any star with a large δc_1 has a larger-than-average δm_1. Also, those stars with" more "negative δm_1 (the Am-type stars) have smaller-than-avergage δc_1. The

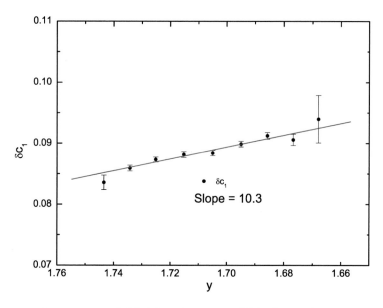

Figure 3: δc_1 versus y divided into 9 groups. This shows the agreement with the expected value of a slope of 9.0 for A-type stars derived for different stars by Crawford. The agreement can be improved by taking the mass effect into account.

values of both indices for FG Vir are located in a very "normal range. In Fig. 5 of Crawford's paper a trend can be found. Fig. 4 shows the relation of δc_1 and δm_1 for FG Vir. A linear fit to the FG Vir data in Fig. 4 lies completely within the cloud of points shown by Crawford (1979) in Fig. 5 of his paper. There is a clear trend in the sense of larger δc_1 values as the δm_1 values are also larger. This is the same trend as in the Crawford paper, but it is evident that the interpretation must be quite different (if existing) for a pulsating star. What does this trend mean in the case of FG Vir? If we inspect Fig. 2, $\delta m_1 = \delta m_1(\beta)$, we can either assume whether the "central" cloud of points (centred at $\beta = 2.785$ mag) is "only" due to "dispersal" of the data points or whether it is related to the pulsation phase of the star. And there are some more questions addressing the relation between the investigated properties and the phase. Questions with this respect are not subject of the current paper. FG Vir is a multiperiodic star and thus makes investigations very complex.

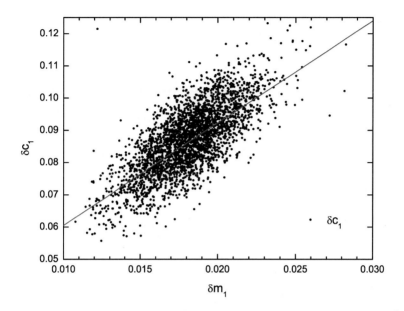

Figure 4: Relation between δc_1 and δm_1 for the measured values

6. Testing the temperature dependence of $M_V(\beta)$

In this section the trend between the colour measurements $y(\beta)$ of FG Vir and the calculated values for the absolute magnitude $M_V(\beta)$ according to Crawford's (1979) calibration are compared. This calibration determines $M_V(\beta)$ by means of $M_V(\mathrm{ZAMS},\ \beta)$ denoting the stars at the position of the ZAMS and $\delta c_1(\beta)$ according to:

$$M_V(\beta) = M_V(\mathrm{ZAMS},\ \beta) - f.\delta c_1(\beta) .$$

$M_V(\mathrm{ZAMS},\ \beta)$ and $c_1(\mathrm{ZAMS},\ \beta)$ are taken from Table I in Crawford's (1979) paper. Crawford proposes $f = 9$ for A-type stars and

$$\delta c_1(\beta) = c_1(\beta) - c_1(\mathrm{ZAMS},\ \beta) .$$

Hypothesis: the trend of $M_V(\beta)$, determined by Crawford's (1979) calibration as outlined above, confirms the trend of the measurements of $y(\beta)$ for FG Vir, i.e., M_V fits the measurements of $y = V$.

To verify this hypothesis the calculated values of M_V are fitted linearly to obtain the trend for $M_V(\beta)$, which is then compared to the trend of the measurements of $y(\beta)$ for FG Vir (also based on a linear fit).

Result: the initially postulated hypothesis for the trend of $M_V(\beta)$ cannot be confirmed under the strict application of Crawford's (1979, 1987) calibration procedure and Crawford's data for the position of the ZAMS. The fit between $M_V(\beta)$ and $y(\beta)$ of FG Vir is unsatisfactory.

The poor agreement between both trends can be seen immediately in Fig. 5.

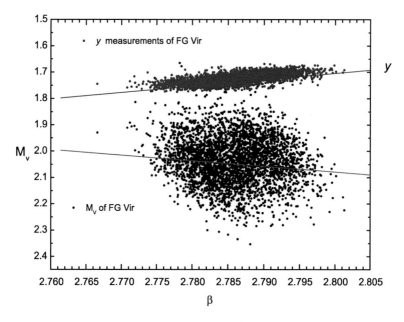

Figure 5: Comparison of the trend line of $M_V(\beta)$ according to Crawford's calibration and the measured values of y(β).

I examined which changes to the calibration procedure are necessary to improve the agreement between the trends of $M_V(\beta)$ and $y(\beta)$. The first step was to modify the factor f. Although a reduction of this factor improves the agreement between both fits, an optimal value for f could not be found.

The next question was whether a change to the adopted position of the ZAMS provides a better fit and how we can determine the "real" position of the ZAMS.

Mermilliod (1981) published a table for $M_V(\text{ZAMS}, B - V)$ for the position of the ZAMS in the Johnson *UBV* system and added the remark: "...The present material may be used to redetermine the ZAMS." The table shows a different temperature dependency for M_V(ZAMS) compared to the one published by Crawford (1979). A transformation as indicated by Caldwell (1993) allows the representation of the table M_V(ZAMS, $B - V$) in a form suitable for the Strömgren *uvby*β photometric system. The substitution of Crawford's position of the ZAMS by Mermilliod's position of the ZAMS proved to be very effective for improving the agreement of the trends between $M_V(\beta)$ and $y(\beta)$. For the calibration of the Strömgren parameters Crawford (1979) selected the open clusters α Per, the Pleiades, IC 4665, the Hyades and data from the Coma cluster, UMa and Praesepe. In order to determine the position of the ZAMS, Mermilliod (1981) selected the open clusters α Per, the Pleiades, IC 4665 and additional open clusters younger than the Hyades.

With this modification to the position of the ZAMS the agreement between both fits improves substantially, but is still not perfect.

Fig. 6 illustrates the disagreement between both positions of the ZAMS definitions in the temperature region of FG Vir.

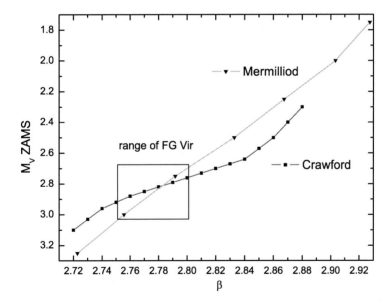

Figure 6: Comparison between the position of the ZAMS of Crawford and the position of the ZAMS of Mermilliod. The rectangle marks the temperature region of FG Vir.

We find that Mermilliod's values for the M_V values of the position of the ZAMS, $M_V(\text{Mer}, \beta)$, and a reduction of the factor f in the calibration formula provides a better fit for M_V of FG Vir. This allows us to draw the following conclusion:

The optimal calibration leading to the best agreement between the trend of the measured values $y(\beta)$ and the trend of the calibrated $M_V(\beta)$ for FG Vir is obtained by the following equation (see Fig. 7):

$$M_V = M_V(\text{Mer}, \beta) - 7.1\delta c_1 .$$

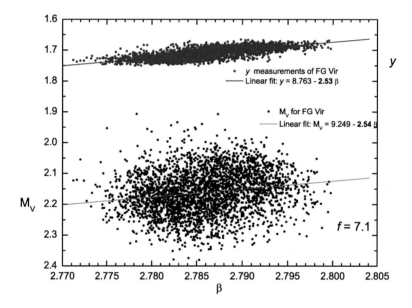

Figure 7: Comparison of the trend line for the measured values of $y(\beta)$ and trend line of $M_V(\beta)$ calibrated with $M_V(\text{ZAMS}, \beta)$ as indicated by Mermilliod and the factor f reduced to $f = 7.1$.

These results, however, are still tentative since the effect of the mass has still to be taken into account (see section 7.). Hence, the next step will be to consider the effect of different masses. Are the different values for the factor f in Crawford's (1979) calibration and in the optimal calibration obtained in this paper a result of mass differences?

7. Effect of mass on the fundamental parameters of a main sequence star

In this paragraph the effect of the mass difference between FG Vir and main sequence stars with the same luminosity and temperature on $\log g$ and c_1 will be investigated.

Stars located on the main sequence are in similar stages of their evolution and therefore exhibit a similar internal structure. Therefore, a mass, $\mathcal{M}$, is in accordance with a specific absolute bolometric magnitude M_{bol} and hence a corresponding luminosity L. In Crawford's (1979) calibration procedure this relationship was kept in mind for A-type stars implicitly for the absolute magnitude M_V (and with it the determination of L and M_{bol}). The analysis of my paper is focused on the application of the calibration relation of Strömgren photometry to determine the luminosity L of the single star FG Vir, which does not change its mass $\mathcal{M}$. During the pulsation cycle of the star mainly the variations of the fundamental parameters T_{eff} (included in $b - y$ and β), $\log g$ (included in c_1), L (included in y or in M_V, determined according to the calibration of Crawford) as well as the radius R are analysed. The mass $\mathcal{M}$ of FG Vir will remain unchanged during the pulsational variations and hence causes a different constellation in comparison with a normal main sequence star (this can also be seen as a perturbation of the equilibrium of the star). One can also make a prediction which mass $\mathcal{M}$ is consistent with the respective luminosity L for every pulsation phase and hence find out the difference of the mass of FG Vir compared to a "nominal value" (or position of an equilibrium model star). In other words, which should be the mass of the star FG Vir in every phase of the pulsation cycle in comparison with an unperturbed star?

I made quantitative estimates of the mass differences and their effect especially on the values of $\log g$ based on the same models and software as used by Lenz et al. (2008). Metallicity and rotational velocity of the stars are selected in accordance with FG Vir. The evolutionary tracks in Fig. 8 for different masses allow an estimate of the difference between the mass of FG Vir and the "nominal mass" of a corresponding unperturbed main sequence star. During each pulsation phase $\log L$ of FG Vir has to be identical to $\log L$ of a corresponding main sequence star. Furthermore, by definition the surface temperature T_{eff} is also the same for both stars. Taking into account the above constraints based on the relations between the fundamental parameters of a star, the following relation between surface gravity g and mass $\mathcal{M}$ can be derived:

$$g_{FG}/g_{MS} = \mathcal{M}_{FG}/\mathcal{M}_{MS} ,$$

where the subscript FG denotes FG Vir and MS a corresponding main sequence star. The deviation $(g_{FG} - g_{MS})$ caused by the mass difference between

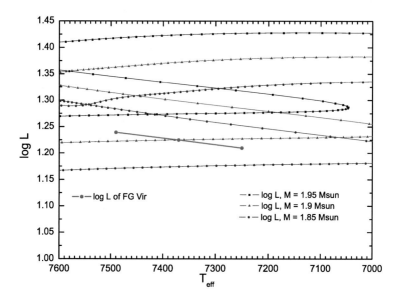

Figure 8: Comparison of evolutionary tracks for 3 main sequence stars with 3 different masses and a corresponding curve spanned by FG Vir with more than 75 frequencies ($\log L = \log L/L_{\odot}$).

FG Vir and a main sequence star with identical luminosity and temperature has an equivalent δc_1 value of $(c_{1FG} - c_{1MS})$. By means of the $c_1(\beta)$ diagram the corresponding c_1 values of the respective main sequence star can be determined. Metallicity and microturbulence of the stars in the diagram are selected in accordance with FG Vir. A description of the used atmosphere model grids can be found in Nendwich et al. (2004) and Heiter et al. (2002). Table 2 highlights 3 points of the temperature range (mean, maximum and minimum) caused by the pulsational variation range of FG Vir comparing the measured value of c_{1FG} to the respective "nominal value" c_{1MS} of the corresponding main sequence star. By definition the deviation of the mean value is chosen to be 0.0.

In the range of the temperature variation of FG Vir the maximal deviation of the nominal value c_{1MS} of the corresponding main sequence star is less than 5.5 % . One can substitute FG Vir's c_{1FG} by c_{1MS} of the corresponding main sequence star and compare the trends of the measured y values and M_V calculated according to Crawford's procedure, for FG Vir with the ZAMS

Table 2: Selected points of the deviation for c_1 between FG Vir and a main sequence star

T_{eff}	β	c_{1FG}	c_{1MS}	$c_{1FG} - c_{1MS}$
7490	2.800	0.860232	0.861322	0.00109
7370	2.786	0.840347	0.840347	0.00000
7250	2.771	0.819042	0.817962	-0.00108

of Mermilliod in place of Crawford's ZAMS. This leads to a good agreement between both trends.

A linear function can be derived from Table 2 for the determination of c_{1MS} as soon as the "measured" c_{1FG} value is known.

8. Conclusions

The current work presents an analysis of the behaviour of the two Strömgren indices c_1 and m_1 based on 3068 records of *uvby* Strömgren photometric data for the δ Scuti star FG Vir. Furthermore, the applicability of Crawford's (1979) calibration procedure for the determination of the trend of the absolute magnitude M_V during pulsations is critically tested by comparing the M_V calibration of FG Vir with the corresponding observed values.

The testing of the trend of m_1 of FG Vir confirms that the metallicity is not varying during the pulsation cycle. It fits with the trend of the standard stars. This can be seen by the fact that the blanketing parameter δm_1, which characterizes the metal content of the star, remains constant. Additionally to previous results for High Amplitude δ Scuti stars (Rodríguez et al. 1991), the confirmation is possible with the given data even for a δ Scuti star with low amplitude.

The testing of the trend of the absolute magnitude M_V, determined according to Crawford's (1979) calibration procedure, shows a poor agreement with the trend of the measured intrinsic colour y. The published calibration of M_V does not fit the measured values of FG Vir. I outline an approach to reach a satisfactory agreement by taking into account the effect of different masses on the trends of log g and c_1, especially by substituting the position of the ZAMS as outlined by Crawford (1979) with the position of the ZAMS as defined by Mermilliod (1981). Mermilliod indicates that a redefinition of the position of the ZAMS will be necessary based on his observations. The current work fully supports this statement.

Acknowledgments. The author thanks Michel Breger for suggesting the subject of this research, for guiding the progress of the research, and for giving invaluable assistance and suggestions during the initial and final analysis of the data in this investigation.

The *uvby* data were collected at SNO (Sierra Nevada Observatory).

This work was supported by the Austrian Fonds zur Förderung der wissenschaftlichen Forschung (FWF).

References

Breger, M., Rodler, F., Pretorius, M. L., et al. 2004, A&A, 419, 695

Breger, M., Lenz, P., Antoci, V., et al. 2005, A&A, 435, 955

Caldwell, J. A. R., Cousins, A. W. J., Ahlers, C. C., et al. 1993, SAAO, 15, 1

Crawford, D. L. 1966, IAUS, 24, 170

Crawford, D. L. 1975, AJ, 80, 955

Crawford, D. L. 1978, AJ, 83, 48

Crawford, D. L. 1979, AJ, 84, 1858

Heiter, U., Kupka, F., van 't Ver-Mennert, C., et al. 2002, A&A, 392, 619

Lenz, P., Pamyatnykh, A. A., Breger, M., & Antoci, V. 2008, A&A, 478, 855

Lester, J. B., Gray, R. O., & Kurucz, R. L. 1986, ApJS, 61, 509

Mermilliod, J. C. 1981, A&A, 97, 235

Moon, T. T., & Dworetsky, M. M. 1985, MNRAS, 217, 305

Nendwich, J., Heiter, U., Kupka, F., et al. 2004, CoAst, 144, 43

Nissen, P. E., Twarog, B. A., & Crawford, D. L. 1987, AJ, 93, 634

Önehag, A., Gustafsson, B., Eriksson, K., & Edvardsson, B. 2009, A&A, 498, 527

Perry, C. L., Olsen, E. H., & Crawford, D. L. 1987, PASP, 99, 1184

Philip A.G.D., & Egret, D. 1980, A&AS 40, 199

Rodríguez, E., Rolland, A., López de Coca, P., & Garrido, R. 1991, A&A, 247, 77

Rodríguez, E., Rolland, A., López de Coca, P., et al. 1992, A&AS, 96, 429

Rodríguez, E., Rolland, A., & López de Coca, P. 1993, A&AS, 101, 421

Rodríguez, E., Rolland, A., Garrido, R., et al. 1998, A&A, 331, 171

Rodríguez, E., López-González, M. J., P., & López de Coca, P. 2000, A&AS, 144, 469

Rodríguez, E., Arellano Ferro, A., Costa, V., et al. 2003, A&A, 407, 1059

Rolland, A., Rodríguez, E., López de Coca, P., et. al. 1991, A&AS, 91, 347

Smalley, B., & Dworetsky, M. M. 1993, A&A, 271, 515

Smalley, B. 1993, A&A, 273, 391

Stankov, A., Sinachopoulos, D. A., Elst, E., & Breger, M. 2002, CoAst, 141, 72

Strömgren, B. 1963, QJRAS, 4, 8

Strömgren, B. 1966, ARA&A, 4, 433

Comm. in Asteroseismology
Volume 162, February 2011
© Austrian Academy of Sciences

A spectroscopic study of the hybrid pulsator γ Pegasi[*]

C. P. Pandey[1,2], T. Morel[3], M. Briquet[4],
K. Jayakumar[5], S. Bisht[1], and B. B. Sanwal[2]

[1] Department of Physics, Kumaun University Nainital - 263002, India.
[2] Aryabhatta Research Institute of Observational Sciences, Nainital - 263129, India.
[3] Institut d'Astrophysique et de Géophysique, Université de Liège, Allée du 6 Août,
Bât. B5c, 4000 Liège, Belgium.
[4] Katholieke Universiteit Leuven, Departement Natuurkunde en Sterrenkunde,
Instituut voor Sterrenkunde, Celestijnenlaan 200D, B-3001 Leuven, Belgium.
[5] Vainu Bappu Observatory, Indian Institute of Astrophysics, Kavalur - 635701, India.

Abstract

The recent detection of both pressure and high-order gravity modes in the classical B-type pulsator γ Pegasi offers promising prospects for probing its internal structure through seismic studies. To aid further modelling of this star, we present the results of a detailed NLTE abundance analysis based on a large number of time-resolved, high-quality spectra. A chemical composition typical of nearby B-type stars is found. The hybrid nature of this star is consistent with its location in the overlapping region of the instability strips for β Cephei and slowly pulsating B stars computed using OP opacity tables, although OPAL calculations may also be compatible with the observations once the uncertainties in the stellar parameters and the current limitations of the stability calculations are taken into account. The two known frequencies $f_1 = 6.58974$ and $f_2 = 0.68241$ d^{-1} are detected in the spectroscopic time series. A mode identification is attempted for the low-frequency signal, which can be associated to a high-order g-mode. Finally, we re-assess the binary status of γ Peg and find no evidence for variations that can be ascribed to orbital motion, contrary to previous claims in the literature.

Accepted: February 16, 2011

Individual Objects: γ Pegasi.

---

[*]Based on observations obtained at Vainu Bappu Observatory, Kavalur, India.

1. Introduction

Of particular relevance for our understanding of the fundamental properties of stars on the upper main sequence are the so-called hybrid B pulsators, which simultaneously exhibit low-order pressure and high-order gravity modes characteristic of β Cephei and slowly pulsating B stars (hereafter SPBs), respectively (in a similar vein, some A- and F-type main sequence stars also present γ Doradus- and δ Scuti-like pulsations; e.g., Grigahcène et al. 2010). A number of hybrid β Cephei/SPB pulsating candidates have been identified to date (e.g., De Cat et al. 2007, Pigulski & Pojmański 2008, Degroote et al. 2009), but their number is expected to grow dramatically in the future as intensive space observations with unprecedented duty cycle and photometric precision are being undertaken (e.g., Balona et al. 2011). The self-driven excitation of both pressure and gravity modes holds great asteroseismic potential because it offers an opportunity to probe both the stellar envelope and the deep internal layers. Such stars are hence prime targets for in-depth seismic modelling and, although challenging, their study promises to lead to significant advances in our understanding of the internal structure of main-sequence B stars (e.g., internal rotation profile, extent of the convective core; Thoul 2009).

Of particular interest in this context is the bright B2 IV star γ Peg (HR 39, HD 886). Although for long considered as a classical β Cephei star, it has recently also been shown to exhibit gravity mode pulsations typical of SPBs, as first shown by Chapellier et al. (2006) using ground-based spectroscopic observations. This result has recently been confirmed and extended by Handler et al. (2009) from high-precision photometry with the *MOST* satellite and a coordinated radial-velocity monitoring from the ground, with the detection of eight β Cep-like and six SPB-like oscillation modes (see also Handler 2009). With this peculiarity, γ Peg enters the (so far) restricted club of hybrid pulsators and is expected to play in the future a pivotal role towards a better understanding of the physics of B stars. Modelling this star also has the potential to address even more fundamental issues, such as the reliability of opacity calculations (see the attempts in this direction by Walczak & Daszyńska-Daszkiewicz 2010 or Zdravkov & Pamyatnykh 2009).

However, the reliability of the results provided by such seismic studies strongly hinges upon an accurate knowledge of both the fundamental parameters of the star under study (e.g., Teff) and its metal mixture. For this reason, an accurate determination of these quantities is of vital importance. In virtue of its brightness and initial status as a prototypical β Cephei star, many abundance works have been devoted to γ Peg (e.g., the pioneering work of Aller 1949), but the analyses very often relied on non fully line-blanketed model atmospheres or fundamental parameters inferred from photometric calibrations and/or LTE calculations (e.g., Gies & Lambert 1992, Pintado & Adelman 1993, Ryans et al. 1996). These limiting assumptions cast some doubts on the reliability of these results. For instance, fitting of the wings of the Balmer lines using LTE synthetic spectra have been shown to systematically lead to an overestimation of the surface gravity (Nieva & Przybilla 2007). Only a few studies have been conducted from an NLTE perspective (Andrievsky et al. 1999, Korotin et al. 1999a,b, Morel et al. 2006), all but one of them only deriving the CNO abundances.[1] Here we present a detailed NLTE abundance analysis of this star in an effort to re-examine its fundamental properties and its position relative to the theoretical instability domains for SPBs and β Cephei stars. We also take advantage of the large number of time-resolved, high-quality spectra collected to re-assess its binarity and to attempt to identify the most prominent modes visible in spectroscopy.

2. Observations and data reduction

The spectroscopic observations were obtained at Vainu Bappu Observatory (VBO) located in Kavalur (India) using the fiber-fed échelle spectrograph attached to the prime focus of the 2.3-m Vainu Bappu telescope (VBT). The wavelength range was 4000-8000 Å for the 2 K $\times$ 4 K CCD detector (spread over 45 orders) and 4200-7000 Å for the 1 K $\times$ 1 K CCD detector (spread over nearly 25 orders with the échelle gaps). The resolving power estimated from the arc spectra is $R = \lambda/\Delta\lambda \approx 60\,000$. Further details regarding the instrumental set up are given in Rao et al. (2005).

A total of 163 spectra were collected from September 2007 to January 2009 during various observing runs. The journal of observations is presented in Table 1. The exposure time ranged from 10 to 15 minutes depending on the sky conditions (i.e., clear sky or sky with thin clouds) and the signal-to-noise ratio was generally above 200. The data were reduced and analysed using standard IRAF (Image Reduction and Analysis Facility)[2] tasks. The basic steps of the data reduction included trimming, bias frame subtraction, scattered light removal, flat fielding, order extraction, wavelength calibration (using ThAr lamps) and finally continuum rectification.

---

[1] Note that the NLTE abundances of Gies & Lambert (1992) are not based on full NLTE line-formation calculations, but are derived instead from the LTE values assuming theoretical NLTE corrections appropriate for a star with parameters typical of γ Peg.

[2] IRAF is distributed by the National Optical Astronomy Observatories, operated by the Association of Universities for Research in Astronomy, Inc., under cooperative agreement with the National Science Foundation.

Table 1: Journal of observations.

Civil Date	Number of spectra	Civil Date	Number of spectra
20 Sep. 07	10	02 Feb. 08	4
28 Sep. 07	19	03 Jun. 08	4
02 Oct. 07	24	04 Jun. 08	4
04 Oct. 07	11	05 Jun. 08	2
05 Oct. 07	28	24 Aug. 08	5
01 Jan. 08	3	31 Aug. 08	3
02 Jan. 08	11	04 Oct. 08	7
03 Jan. 08	5	05 Oct. 08	11
16 Jan. 08	3	13 Jan. 09	3
17 Jan. 08	3	14 Jan. 09	3
		Total	163

3.   Line-profile variations

It should be noted that because of different instrumental settings, not all spectral lines were systematically covered during the observations. We therefore focus in the following on the strong spectral lines that were the most extensively observed, namely the Si III triplet between 4552 and 4575 Å and the C II $\lambda\lambda$5143, 5145 doublet, with 80 and 131 exposures, respectively.

3.1.   Binarity

Until recently, γ Peg was believed to be a spectroscopic binary whose orbital period was, however, disputed ($P_{orb} = 370.5$ days, Chapellier et al. 2006; $P_{orb} = 6.816$ days, Harmanec et al. 1979, Butkovskaya & Plachinda 2007). Contrary to previous authors, Handler et al. (2009) suggested γ Peg to be a single star and explained the claimed orbital variations as due to g-mode pulsation.

 To re-assess the possible binarity of this object, we compared our radial velocities (RVs) of the C II λ5145 line with the orbital solution proposed by Chapellier et al. (2006). Our heliocentrically-corrected RVs vary between -1.5 and 7 km s^{-1} (with a typical accuracy on the individual measurement of $\sim$0.3 km s^{-1}) mainly as a result of the well-known, dominant radial mode with a period of about 0.15 day (see below). There is no evidence for any long-term trends, despite the fact that our observations cover some phase intervals (most notably around $\phi\sim$0.93) where rapid variations amounting to up to $\sim$12 km s^{-1} are expected according to the ephemeris of Chapellier et al. (2006). This rules out the possibility that γ Peg is an eccentric binary with an orbital period of about one year, confirming the conclusion of Handler et al. (2009). On the other hand, these authors favoured the pulsation interpretation for the ~6.8 days period since they noticed that the frequency 1/6.816 d^{-1} with an amplitude of ~0.8 km s^{-1} (Butkovskaya & Plachinda 2007), is a one-day alias of the frequency 0.8533 d^{-1}, which lies well within the domain of g-mode frequencies. Finally, our data also do not support the large, abrupt RV changes reported by Butkovskaya & Plachinda (2007).

3.2.   Frequency analysis and mode identification

To further investigate the line-profile variability in γ Peg, we used the software package FAMIAS[3] (Zima 2008). We computed the first three velocity moments $< v^1 >$, $< v^2 >$ and $< v^3 >$ (see Aerts et al. 1992 for a definition of the moments of a line profile) of the Si III triplet and the C II doublet with the aim of performing a frequency analysis. The integration limits for computing the moments were dynamically chosen by sigma clipping to avoid the noisy continuum.

In the first moment $< v^1 >$, which is the RV placed at average zero, we detected the well-known dominant radial mode, which can readily be seen in the line-profile variations affecting the Si III lines (Fig. 1), followed by a one-day alias of the known highest amplitude low frequency mode. Both pulsation modes were first discovered by Chapellier et al. (2006) and afterwards confirmed by Handler et al. (2009), who definitely proved the hybrid β Cep/SPB nature of γ Peg. Since the frequency values of Handler et al. (2009) are more accurate than ours, we adopted them in our study, i.e., $f_1 = 6.58974$ and $f_2 = 0.68241$ d^{-1}. The corresponding amplitudes in our dataset are 3.39 and 0.69 km s^{-1} with an error of 0.07 km s^{-1}, for f_1 and f_2, respectively. No additional frequencies could be detected according to the 4 signal-to-noise criterion of Breger et al. (1993). The latter was tested by computing the noise level for different box intervals (between 1 and 10 d^{-1}) centred on the considered frequency.

In an attempt to detect other periodicities, we performed a frequency search on the spectra by means of a two-dimensional Fourier analysis. In this way, we again detected f_1 and f_2 but no additional frequencies. A non-linear multiperiodic leastsquares fit of a sum of sinusoidals is then computed with the Levenberg-Marquardt algorithm (adapted from Press et al. 2007). This fitting is applied for every bin of the spectrum separately according to the formula

---

[3]FAMIAS has been developed in the framework of the FP6 European Coordination Action HELAS – http://www.helas-eu.org/

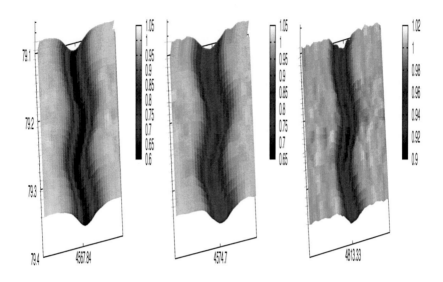

Figure 1: Examples of line-profile variations affecting Si III λ4567.8, 4574.7 and 4813.3 on 5 October 2007.

$Z + \sum_i A_i \sin[2\pi(f_i t + \phi_i)]$, where Z is the zeropoint, and A_i, f_i, and ϕ_i are the amplitude, frequency, and phase of the i-th frequency, respectively. The amplitude and phase distributions across the C II λ5145 Å line are shown in Fig. 2.

To identify the modes associated to f_1 and f_2, we used the Fourier parameter fit method (FPF method; Zima 2006). The wavenumbers (ℓ, m) and other continuous parameters are determined in such a way that the theoretical zeropoint, amplitude and phase values across the profile best fit the observed values. The fitting is carried out by applying genetic optimization routines in a large parameter space. From multicolour photometric time series, Handler (2009) unambiguously identified f_1 as a radial mode while two possibilities remain for f_2, which is a $\ell_2 = 1$ or a $\ell_2 = 2$ mode. Our spectroscopy corroborates that $(\ell_1, m_1) = (0, 0)$, as illustrated in Fig. 2. To identify the values of (ℓ_2, m_2), we only allowed ℓ_2 to be 1 or 2. Such an approach, which consists in adopting the ℓ-values obtained from photometry, already proved to be successful for spectroscopic mode identification of other β Cephei stars (Briquet et al. 2005; Mazumdar et al. 2006; Desmet et al. 2009). In Table 2 the best pa-

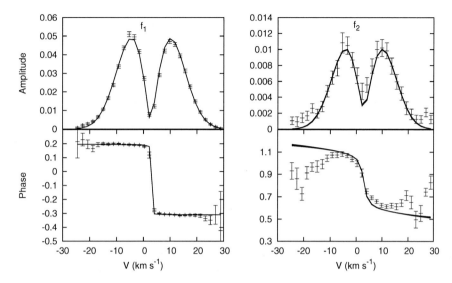

Figure 2: Amplitude and phase distributions (points with error bars) across the C II λ5145 Å line, for the frequencies $f_1 = 6.58974$ and $f_2 = 0.68241$ d^{-1}. The bestfit model is represented by full lines. For f_1, the mode is radial. For f_2, the differences between the best solutions listed in Table 2 for $(\ell_2, m_2) = (1,1)$, $(2, -1)$ and $(2, -2)$ are smaller than the thickness of the lines. The amplitudes are expressed in units of continuum and the phases in π radians.

rameter combinations are listed for each couple (ℓ_2, m_2). The smaller the χ^2 value, the better the solution. The most probable and equally good solutions are $(\ell_2, m_2) = (1,1)$, $(2, -2)$ or $(2, -1)$ (where a positive m-value denotes a prograde mode). The best models for the amplitude and phase across the line profile are illustrated in Fig. 2. A by-product of the mode identification is the derivation of the equatorial rotational velocity v_{eq}. By using the χ^2-values as weights, we constructed a histogram for v_{eq} (see Fig. 3), as in Desmet et al. (2009). We computed it by considering the solutions with $(\ell_2, m_2) = (1,1)$, $(2, -2)$ and $(2, -1)$. By calculating a weighted mean and standard deviation, we obtained $v_{\text{eq}} = 6 \pm 1$ km s^{-1} (a much less probable value is $\sim$12 km s^{-1}).

Table 2: Mode parameters derived from the FPF mono-mode method for f_2. a_s is the surface velocity amplitude (km s^{-1}); i is the stellar inclination angle in degrees; $v \sin i$ is the projected rotational velocity, σ is the width of the intrinsic profile, both expressed in km s^{-1}.

χ^2	(ℓ_2, m_2)	a_s	i	$v \sin i$	σ
3.59	(1,1)	3.9	14	1.3	6.9
3.60	(2, −2)	2.5	25	2.6	6.5
3.62	(2, −1)	2.5	84	6.2	6.1
4.77	(1,0)	6.2	86	5.8	5.3
4.89	(0,0)	2.8	—	1.0	6.9
4.91	(2,0)	0.4	26	4.7	6.6
5.29	(2,1)	5.1	87	1.0	6.4
5.30	(1, −1)	0.9	63	1.0	7.1
5.84	(2,2)	7.4	15	1.0	6.1

4.   Abundance analysis

4.1.   Analysis tools

The atmospheric model calculations were performed under the assumption of LTE, whereas a full NLTE treatment was adopted for the line formation. Such a hybrid approach has been shown to be adequate for early B-type stars on the main sequence (Nieva & Przybilla 2007). First, the ATLAS9 code[4] (Kurucz 1993) is used to compute the hydrostatic, plane-parallel and fully line-blanketed LTE atmospheric models. Grids with the new opacity distribution functions (ODFs) assuming the solar abundances of Grevesse & Sauval (1998) have been used (Castelli & Kurucz 2004).

Further, in order to obtain the NLTE abundances we made use of the latest version of the NLTE line-formation codes DETAIL/SURFACE (Butler & Giddings 1985, Giddings 1981). DETAIL provides the solution of the radiative transfer and statistical equilibrium equations, while the emergent spectrum is calculated by SURFACE. The line atomic data are taken from the NIST and VALD databases. Care has been taken to only retain features that are unblended in the relevant temperature range.

---

[4]http://kurucz.harvard.edu/

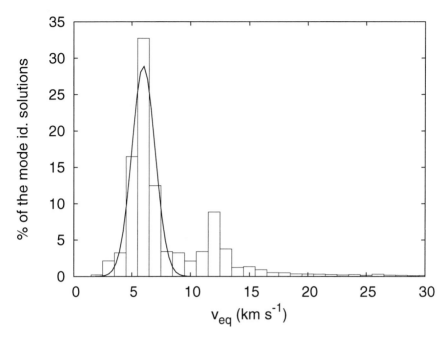

Figure 3: Histogram for the equatorial rotational velocity derived from the FPF method.

4.2.   Determination of the atmospheric parameters

Our abundance analysis is based on the average of a large number of time-resolved spectra (selected to have the highest S/N) to ensure that the parameters and abundances we derive are representative of the values averaged over the whole pulsation cycle.

The effective temperature, T_{eff}, was estimated from the Si II/III ionisation balance. We cannot rely upon Si III/IV ionisation balance, as no Si IV lines are visible. On the other hand, in this T_{eff} range no other chemical elements have lines of two adjacent ionisation stages that can be measured. Numerous Fe III lines and two very weak Fe II lines (Fe II λ5018 and Fe II λ5169) with an EW in the range 10-15 mÅ are present in our spectra. However, constraining T_{eff} from ionisation balance of iron was not possible owing to the rudimentary nature of our Fe II model atom, which only includes 8 levels. In contrast, the detailed treatment of the Fe III ion (264 levels) allows us to reliably estimate the iron abundance from the analysis of the Fe III features. Figure 4 shows examples of calibrations between the Si line ratios and T_{eff}. Ten line ratios have been used for the temperature estimation and the dispersion between the various T_{eff} values obtained was considered as representative of the uncertainty in this parameter.

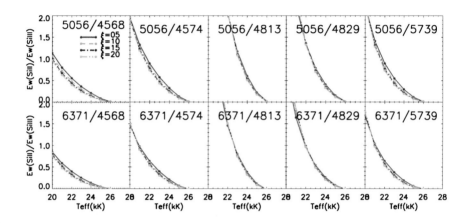

Figure 4: Examples of calibrations between various Si line ratios and the effective temperature, as a function of the microturbulence ξ (red lines: $\xi = 5$ km s^{-1}; light green lines: $\xi = 10$ km s^{-1}; blue lines: $\xi = 15$ km s^{-1}; dark green lines: $\xi = 20$ km s^{-1}). We have adopted $\log g = 3.7$ and $\xi = 5$ km s^{-1}, as appropriate for this star (see text).

Unfortunately, we were unable to estimate the surface gravity, $\log g$, from the fitting of the collisionally-broadened wings of the Balmer lines, as no profiles were completely covered because of échelle gaps. Therefore, this parameter was estimated from Strömgren $uvby\beta$ photometry based on the calibrations of Castelli (1991). The c_1 and β colour indices were taken from the Catalogue of $uvby\beta$ data of Hauck & Mermilliod (1998). The observed c_1 colour index was dereddened to obtain c_0 by means of empirical calibrations (details can be found in Castelli 1991 and references therein). Slightly different sets of colour indices (c_1, β) are available in the literature (Crawford et al. 1971, Sterken & Jerzykiewicz 1993). The surface gravity was computed using all the possible combinations and the scatter was taken as the typical error bar. The simultaneous estimation of the effective temperature and surface gravity, along with the error limits, is illustrated in Fig. 5.

Another important parameter which is required for the abundance analysis is the microturbulent velocity, ξ, which was determined from the O II features requiring that the abundances are independent of the line strength (Fig. 6). The uncertainty in the microturbulence was estimated by varying this parameter until

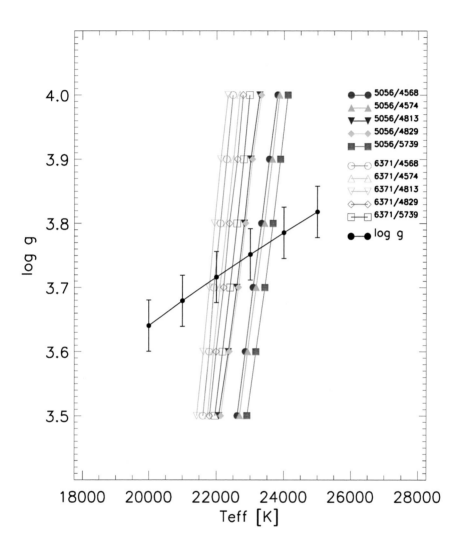

Figure 5: T_{eff}-log g plane for the simultaneous determination of the effective temperature and surface gravity. The group of nearly vertical lines represents the loci satisfying the Si II/Si III ionisation equilibria, while the thick solid line represents the log g determined from Strömgren photometry. The middle point of the intersection of the two sets of lines simultaneously provides the value of T_{eff} and log g.

the slope of the $\log \epsilon(O)$-$\log \epsilon(EW/\lambda)$ differs from zero at the 3σ level.

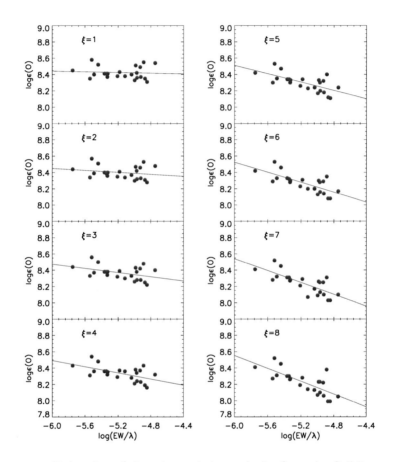

Figure 6: Estimation of the microturbulent velocity from the O II lines.

Finally, the total amount of line broadening, v_T, is determined from line-profile fitting of a number of isolated metallic lines (instrumental broadening was estimated from the calibration lamps). Although rotational broadening clearly dominates over pulsational broadening in this star whose dominant pulsation mode is radial, the fact that this estimate ($v_T = 8\pm1$ km s^{-1}) is marginally higher than the equatorial rotational velocity inferred from the moment analysis ($v_{eq} = 6\pm1$ km s^{-1}) suggests that broadening arising from non-radial pulsations may not be completely negligible. A comparison between the observed and fitted profiles is presented in the Appendix.

The atmospheric parameters, along with their uncertainties, are presented in Table 3. A detailed comparison with previous results in the literature is deferred to the last section.

Table 3: Derived atmospheric parameters for γ Peg.

T_{eff} (K)	22650±650
log T_{eff}	4.355±0.013
log g	3.73±0.08
ξ (km s^{-1})	1^{+2}_{-1}
v_T (km s^{-1})	8±1

4.3. Determination of the elemental abundances

The NLTE abundances of He, C, N, O, Ne, Mg, Al, Si and Fe have been determined from a classical curve-of-growth analysis using the EWs measured through direct integration. The mean abundances are presented in Table 4, while the results for each transition considered are given in the Appendix (along with the EW measurements). To estimate the abundance uncertainties, we have first calculated the errors arising from the uncertainties in the atmospheric parameters (i.e., σ_{Teff}, σ_{logg} and σ_{ξ}). They were derived by computing the abundances using models with atmospheric parameters deviating from the nominal values by the uncertainties tabulated in Table 3. We also considered the fact that the determinations of T_{eff} and log g are strongly coupled. Finally, the total uncertainty, σ_{total}, is obtained by quadratically summing up these errors and the line-to-line scatter (σ_{int}).

After completion of the analysis, we became aware of a study (Simón-Díaz 2010) reporting problems with the modelling of some silicon lines using the model atom implemented in the NLTE code FASTWIND (Puls et al. 2005). Because our model atom is similar in many respects (Morel et al. 2006), we have examined to what extent this would affect our determination of the atmospheric parameters and ultimately abundances by repeating the abundance analysis excluding the spectral lines that may not be properly modelled (namely Si II $\lambda5056$ and Si III $\lambda4813$, 4819, 4829). This leads to a slightly lower temperature (see Fig. 5) and surface gravity: T_{eff} = 22300 K and log g = 3.72 (the microturbulence remains unchanged). The abundances determined using these parameters are provided in Table 5. The differences with the mean abundances derived assuming the default parameters (Table 4) remain within 0.1 dex, except in the case of Si where it amounts to about 0.18 dex.

Table 4: Mean NLTE abundances (the number of lines used for each element is given in brackets) with the usual notation assuming $\log \epsilon(H)=12$, and details on the error budget. The last column gives the total abundance uncertainty.

	Mean abundance		σ_{int}	σ_{Teff}	σ_{logg}	σ_ξ	$\sigma_{Teff/logg}$	σ_{total}
				Error estimation				
$\log \epsilon(He)$	10.92	(8)	0.24	0.04	0.02	0.05	0.03	0.25
$\log \epsilon(C)$	8.26	(9)	0.05	0.00	0.00	0.01	0.01	0.05
$\log \epsilon(N)$	7.62	(22)	0.08	0.07	0.03	0.03	0.03	0.12
$\log \epsilon(O)$	8.43	(23)	0.05	0.15	0.06	0.06	0.09	0.20
$\log \epsilon(Ne)$	8.26	(8)	0.05	0.05	0.01	0.01	0.04	0.08
$\log \epsilon(Mg)$	7.60	(1)	—	0.06	0.01	0.16	0.06	0.18
$\log \epsilon(Al)$	6.30	(3)	0.00	0.04	0.00	0.09	0.02	0.10
$\log \epsilon(Si)$	7.14	(8)	0.18	0.06	0.02	0.09	0.03	0.21
$\log \epsilon(Fe)$	7.30	(24)	0.09	0.08	0.05	0.07	0.03	0.15

Table 5: Mean NLTE abundances (the number of lines used for each element is given in brackets) assuming $T_{eff} = 22300$ K, $\log g = 3.72$ and $\xi = 1$ km s^{-1}.

	Mean abundance	
$\log \epsilon(He)$	10.88	(8)
$\log \epsilon(C)$	8.25	(9)
$\log \epsilon(N)$	7.65	(22)
$\log \epsilon(O)$	8.52	(23)
$\log \epsilon(Ne)$	8.24	(8)
$\log \epsilon(Mg)$	7.56	(1)
$\log \epsilon(Al)$	6.32	(3)
$\log \epsilon(Si)$	7.32	(4)
$\log \epsilon(Fe)$	7.30	(24)

5. Discussion

5.1. Chemical composition

Figure 7 shows a comparison between our abundances (Table 4) and previous NLTE results in the literature (Gies & Lambert 1992, Andrievsky et al. 1999, Korotin et al. 1999a,b, Morel et al. 2006, Morel & Butler 2008), as well as the new solar abundances of Asplund et al. (2009). There is an overall satisfactory agreement with these previous studies. However, all these results taken at face value would suggest that γ Peg is metal poor with respect to the Sun, a fact which is obviously contrary to the expectations for a young star in the solar neighbourhood. This is likely connected to the long-standing and more general problem affecting most abundance analyses of OB stars, which yield abundances significantly lower than solar (e.g., Morel 2009 for a review). Although the origin of this problem remains unclear, improvements in the atomic data and determination of the atmospheric parameters have been claimed to solve most of the discrepancy (Przybilla et al. 2008, Simón-Díaz 2010). In view of this rather unsatisfactory situation, a sound assumption may be to use in further theoretical modelling the abundances determined for a small sample of nearby B-type stars by Przybilla et al. (2008) or the solar abundances of Asplund et al. (2009).

However, a robust result is that the chemical composition of γ Peg does not significantly differ from that of early B-type stars analysed using similar techniques, whether they are known as pulsating or not (Morel et al. 2008). The existence of microscopic diffusion processes in hybrid pulsators leading to an accumulation of iron in the driving zone has often been invoked to explain the unexpected excitation of some specific pulsation modes (e.g., Pamyatnykh et al. 2004), but the lack of any abundance peculiarities does not seem at first glance to support this possibility unless one supposes that the changes at the surface remain below the detection limits. On the other hand, the [N/O] and [N/C] logarithmic abundance ratios (-0.81 ± 0.24 and -0.64 ± 0.13 dex, respectively) are identical to within the errors to the solar values (Asplund et al. 2009). Our study therefore confirms that there is no evidence in γ Peg for CNO-processed material dredged up to the surface because of deep mixing. The significant boron depletion (Proffitt & Quigley 2001) suggests, however, that shallow mixing is already taking place in the superficial layers.

5.2. Comparison with theoretical instability strips

As γ Peg is one of the rare stars known to date to present both high-order g modes and low-order p and g modes, it is of interest to examine whether this property is consistent with the theoretical expectations. The position of γ Peg

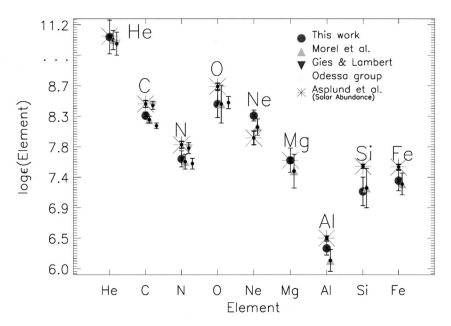

Figure 7: Comparison between our abundances and previous NLTE results in the literature (Gies & Lambert 1992, Morel et al. 2006, Morel & Butler 2008, Andrievsky et al. 1999, Korotin et al. 1999a,b). The data points dubbed 'Odessa group' refer to the results of Andrievsky et al. (1999) and Korotin et al. (1999a,b). The solar abundances of Asplund et al. (2009) are indicated with different symbols.

in the (log T_{eff}, log g) plane is shown in Fig. 8, along with the instability domains for β Cephei and SPB-like pulsation modes computed with the solar abundances of Asplund et al. (2005) and for two different opacity tables, OP and OPAL (see Miglio et al. 2007 for details). At face value, this figure suggests that the hybrid nature of the pulsations in γ Peg is more consistent with instability strips computed with OP opacities, as the OPAL ones do not predict this star to exibit SPB-like pulsations.

As can also be seen in this figure, the same conclusion holds for the two other best-studied hybrid pulsators, ν Eri (Jerzykiewicz et al. 2005) and 12 Lac (Handler et al. 2006), that have been analysed using similar techniques as in the present work (Morel et al. 2006). Identical parameters within the errors were obtained for ν Eri by De Ridder et al. (2004) from a full spectroscopic analysis. Lefever et al. (2010) analysed γ Peg, ν Eri and 12 Lac using the same data as Morel et al. (2006), but with a different code, and also found

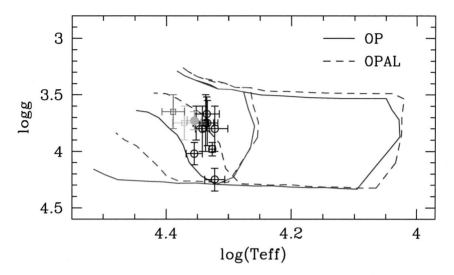

Figure 8: Position of γ Peg in the (log T_{eff}, log g) plane (green filled circle), along with the instability strips for β Cephei- (red lines) and SPB-like (blue lines) pulsation modes ($l < 4$) computed using the solar abundances of Asplund et al. (2005) and the OP and OPAL opacity tables (Miglio et al. 2007). The neon abundance is taken from Cunha et al. (2006). The open circles indicate results for γ Peg in the literature (Fitzpatrick & Massa 2005, Gies & Lambert 1992, Korotin et al. 1999a, Martin 2004, Morel et al. 2006 [based on Si ionisation equilibrium], Morel & Butler 2008 [based on Ne ionisation equilibrium], Niemczura & Połubek 2006, Pintado & Adelman 1993, Ryans et al. 1996). The two open squares refer to the results for two other hybrid pulsators, ν Eri and 12 Lac (Morel et al. 2006).

very similar results, except a temperature lower by 1500 K in 12 Lac. These two studies also determined T_{eff} and log g from Si ionisation balance and fitting of the Balmer lines. On the other hand, and although they may be regarded as less reliable as they heavily rely on only one very weak Ne II line, slightly lower T_{eff} values were obtained for these three stars by Morel & Butler (2008) based on Ne ionisation equilibrium. Finally, Gies & Lambert (1992) found significantly higher T_{eff} values for ν Eri and 12 Lac based on photometric data, but it is likely that their temperature scale is too hot (see discussion in Lyubimkov et al. 2002). This also leads to surface gravities that are higher than other results in the literature.

While the opacity values prove critical for the excitation, or lack thereof, of oscillations in B stars, the location of the SPB instability domain is relatively insensitive to the chemical mixture used (Miglio et al. 2007). Using the recently revised abundances of Asplund et al. (2009) is therefore not expected to change this picture significantly (an identical situation is indeed encountered when using the more different solar mixture of Grevesse & Noels 1993).

To investigate to what extent our conclusions regarding the ability of the opacity calculations to reproduce the observations is robust against the choice of $\log T_{eff}$ and $\log g$, we show in Fig. 8 other results for γ Peg from the literature. Although a variety of techniques have been used for the determination of the parameters, almost all these results rely on photometric data and/or LTE methods. It should also be noted that these estimates are not completely independent, as the same photometric calibrations (and data) have often been employed. Despite these caveats, the gravity seems well constrained to around $\log g \sim 3.8$ if one excludes the discrepant value of Pintado & Adelman (1993) based on LTE fitting of the Hγ line. The effective temperatures are generally based on narrow-band photometric data and appear slightly lower than our value based on Si ionisation equilibrium (keeping in mind that, as discussed above, rejecting some potentially problematic Si lines would lower our T_{eff} by about 350 K). Taking into account the still relatively large inaccuracy of the T_{eff} value, we conclude, solely based on the position of γ Peg with respect to the instability strips, that the OP opacities are only marginally preferred to the OPAL ones. Accurate positions for a sufficiently large sample of stars may lead to stronger statements. It should also be kept in mind, however, that some important physical processes that are not currently implemented in the stability analysis may substantially change the results of the models (e.g., stellar rotation; Townsend 2005).

To conclude, although it is clear that several theoretical aspects related to the excitation/damping of oscillations in B stars may greatly benefit from the constraints offered by the further identification of hybrid pulsators and from a detailed study of their pulsation properties, an accurate determination of their fundamental parameters remains essential for further progress.

Acknowledgments. We are indebted to Dr. K. Butler for his valuable comments and for kindly providing us with the codes DETAIL/SURFACE. C.P.P. sincerely thanks to the members of the technical staff (M. Appakutti, V. Moorthy and C. Velu) at VBO (Kavalur, India) for their kind assistance during the observations, Profs. N. K. Rao, S. Giridhar, H. C. Bhatt, and B. P. Das, Drs. C. Muthumariappan and G. Pandey for their kind hospitality and encouragement during her visits to the Indian Institute of Astrophysics (Bengaluru, India), as well as Dr. U. S. Chaubey for his encouragement and for providing the financial assistance for her visits to VBO, through a sponsored project (Grant No SR/S2/HEP-20/2003) of the Department of Science and Technology (DST), Government of India, New Delhi and also Prof. Ram Sagar for his kind motivation. T. M. acknowledges financial support from Belspo for contract PRODEX-GAIA DPAC. M.B. is Postdoctoral Fellow of the Fund for Scientific Research, Flanders. We are grateful to Andrea Miglio for kindly providing us the theoretical instability strips for B-type pulsators and for useful comments. The authors are also grateful to the anonymous referee for his/her constructive comments which helped in improving the presentation of the paper. This research made use of NASA Astrophysics Data System Bibliographic Services, the SIMBAD database operated at CDS, Strasbourg (France).

References

Aerts, C., de Pauw, M., & Waelkens, C. 1992, A&A, 266, 294

Aller, L. H. 1949, ApJ, 109, 244

Andrievsky, S. M., Korotin, S. A., Luck, R. E., & Kostynchuk, L. Yu. 1999, A&A, 350, 598

Asplund, M., Grevesse, N., & Sauval, A. J. 2005, ASP Conf. Ser., 336, 25

Asplund, M., Grevesse, N., Sauval, A. J., & Scatt, P. 2009, ARA&A 47, 481

Balona, L. A., Pigulski, A., De Cat, P., et al. 2011, MNRAS, in press

Breger, M., Stich, J., Garrido, R., et al. 1993, A&A, 271, 482

Briquet, M., Lefever, K., Uytterhoeven, K., & Aerts, C. 2005, MNRAS, 362, 619

Butkovskaya, V. V., & Plachinda, S. I. 2007, A&A, 469, 1069

Butler, K., & Giddings, J. R. 1985, Newsletter of Analysis of Astronomical Spectra, No.9 (Univ. London)

Castelli, F. 1991, A&A, 251, 106

Castelli, F., & Kurucz, R. L. 2004, IAU Symp., 210, A20 (arXiv:astro-ph/0405087)

Chapellier, E., Le Contel, D., Le Contel, J. M., et al. 2006, A&A, 448, 697

Crawford, D. L., Barnes, J. V., & Golson, J.C. 1971, AJ, 76, 1058

Cunha, K., Hubeny, I., & Lanz, T. 2006, ApJ, 647, 143

De Cat, P., Briquet, M., Aerts, C., et al. 2007, A&A, 463, 243

Degroote, P., Briquet, M., Catala, C., et al. 2009, A&A, 506, 111

De Ridder, J., Telting, J. H., Balona, L. A., et al. 2004, MNRAS, 351, 324

Desmet, M., Briquet, M., Thoul, A., et al. 2009, MNRAS, 396, 1460

Fitzpatrick, E. L., & Massa, D. 2005, AJ, 129, 1642

Giddings, J. R. 1981, Ph.D. Thesis , University of London

Gies, D. R., & Lambert, D. L. 1992, ApJ, 387, 673

Grevesse, N., & Noels, A. 1993, in Hauck B. and Paltani S. R. D., eds, La Formation des Eléments Chimiques. AVCP, Lausanne, 205

Grevesse, N., & Sauval, A. J. 1998, Space Sci. Rev., 85, 161

Grigahcène, A., Antoci, V., Balona, L., et al. 2010, ApJ, 713, L192

Handler, G., Jerzykiewicz, M., Rodriguez, E., et al. 2006, MNRAS, 365, 327

Handler, G. 2009, MNRAS, 398, 1339

Handler, G., Matthews, J. M., Eaton, J. A., et al. 2009, ApJ, 698, 56

Harmanec, P., Koubsky, P., Krpata, J., & Zdarsky, F. 1979, IBVS, 1590

Hauck, B., & Mermilliod, M. 1998, A&AS, 129, 431

Jerzykiewicz, M., Handler, G., Shobbrook, R. R., et al. 2005, MNRAS, 360, 619

Korotin, S. A., Andrievsky, S. M., & Kostynchuk, L. Yu. 1999a, A&A, 342, 756

Korotin, S. A., Andrievsky, S. M., & Luck, R. E. 1999b, A&A, 351, 168

Kurucz, R. L. 1993, ATLAS9 Stellar Atmosphere Programs and 2 km/s grid. Kurucz CD-ROM No. 13. Cambridge, Mass.: Smithsonian Astrophysical Observatory, 13

Lefever, K., Puls, J., Morel, T., et al. 2010, A&A , 515, A74

Lyubimkov, L. S., Rachkovkaya, T. M., Rostopchin, S. I., et al. 2002, MNRAS, 333, 9

Martin, J. C. 2004, AJ, 128, 2474

Mazumdar, A., Briquet, M., Desmet, M., & Aerts, C. 2006, A&A, 459, 589

Miglio, A., Montalbán, J., & Dupret, M.-A. 2007, MNRAS, 375, 21

Morel, T., Butler, K., Aerts, C., et al. 2006, A&A, 457, 651

Morel, T., Hubrig, S., & Briquet, M. 2008, A&A, 481, 453

Morel, T., & Butler, K. 2008, A&A, 487, 307

Morel, T. 2009, CoAst, 158, 122

Niemczura, E., & Połubek, G. 2006, ESASP, 624, 120

Nieva, M. -F., & Przybilla, N. 2007, A&A, 467, 295

Pamyatnykh, A. A., Handler, G., & Dziembowski, W. A. 2004, MNRAS, 350, 1022

Pigulski, A., & Pojmański, G. 2008, A&A, 477, 917

Pintado, O. I., & Adelman, S. J. 1993, MNRAS, 264, 63

Press, W. H., Teukolsky, S. A., Vetterling, W. T., & Flannery, B. P., Numerical Recipes 3rd Edition: The Art of Scientific Computing, Cambridge University Press, New York, 2007

Proffitt, C. R., & Quigley, M. F. 2001, ApJ, 548, 429

Przybilla, N., Nieva, M. F., & Butler, K. 2008, ApJ, 688, 103

Puls, J., Urbaneja, M. A., Venero, R., et al. 2005, A&A, 435, 669

Rao, N. K., Sriram, S., Jayakumar, K., & Gabriel, F. 2005, JAA, 26, 331

Ryans, R. S. I., Hambly, N. C., Dufton, P. L., & Keenan, F. P. 1996, MNRAS, 278, 132

Simón-Díaz, S. 2010, A&A, 510, 22

Sterken, C., & Jerzykiewicz, M. 1993, SSRv, 62,95

Thoul, A. 2009, CoAst, 159,35

Townsend, R. H. D. 2005, MNRAS, 360, 465

Walczak, P., & Daszyńska-Daszkiewicz, J. 2010, AN, 331, 1057

Zdravkov, T., & Pamyatnykh, A. A. 2009, AIPC, 1170, 388

Zima, W. 2006, A&A, 455, 227

Zima, W. 2008, CoAst, 155, 1

Appendix

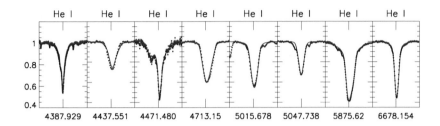

Figure 9a: Comparison between the observed and synthetic profiles (represented by the dots and full lines, respectively) computed for the abundance yielded by the corresponding He I line. The synthetic spectra have been convolved with a rotational broadening function with $v_T=8$ km s^{-1}.

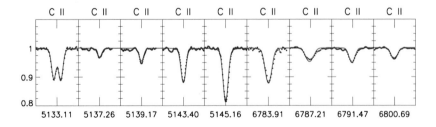

Figure 9b: Same as Fig. 9a, but for the C II lines.

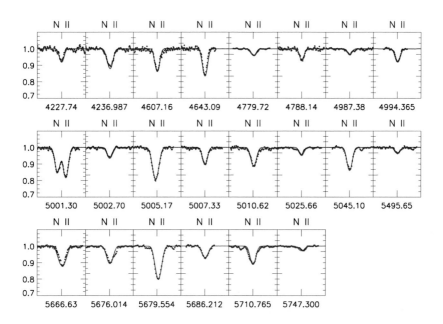

Figure 9c: Same as Fig. 9a, but for the N II lines.

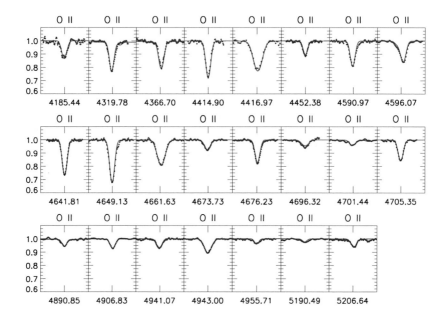

Figure 9d: Same as Fig. 9a, but for the O II lines.

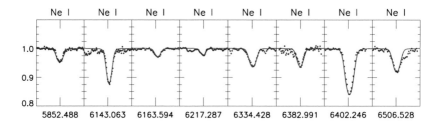

Figure 9e: Same as Fig. 9a, but for the Ne I lines.

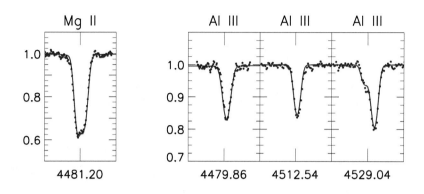

Figure 9f: Same as Fig. 9a, but for the Mg II (left) and Al III (right) lines.

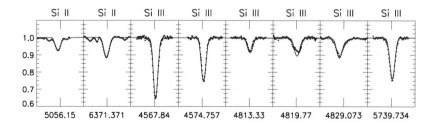

Figure 9g: Same as Fig. 9a, but for the Si II and Si III lines.

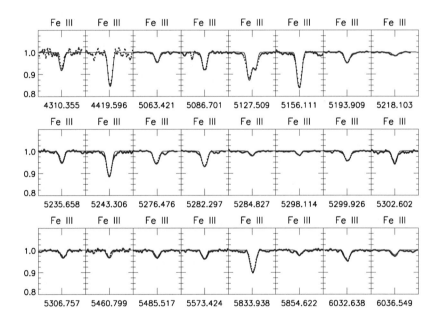

Figure 9h: Same as Fig. 9a, but for the Fe III lines.

Table 6: EW measurements and abundance results for each transition, along with the error estimates.

Ion	Transition (Å)	EW (mÅ)	log ϵ (dex)	σ_{Teff}	σ_{logg}	σ_ξ	$\sigma_{Teff/logg}$	σ_{total}
He I								
	4387.93	820	11.29	0.08	0.03	0.00	0.04	0.09
	4437.55	109	10.98	0.06	0.00	0.06	0.06	0.10
	4471.48	1255	11.09	0.05	0.04	0.00	0.00	0.06
	4713.15	270	10.92	0.07	0.00	0.07	0.07	0.12
	5015.68	274	10.76	0.00	0.09	0.10	0.00	0.13
	5047.74	170	11.04	0.06	0.00	0.06	0.06	0.10
	5875.62	605	10.53	0.00	0.00	0.00	0.00	0.00
	6678.15	560	10.76	0.00	0.00	0.10	0.00	0.10
C II	5133.11	62	8.22	0.01	0.00	0.01	0.01	0.02
	5137.26	9	8.31	0.00	0.00	0.00	0.01	0.01
	5139.17	14	8.33	0.01	0.01	0.01	0.02	0.03
	5143.40	33	8.26	0.01	0.00	0.01	0.01	0.02
	5145.16	58	8.25	0.01	0.01	0.03	0.01	0.03
	6783.91	50	8.25	0.00	0.01	0.03	0.01	0.03
	6787.21	17	8.30	0.01	0.00	0.01	0.01	0.02
	6791.47	18	8.23	0.00	0.01	0.01	0.01	0.02
	6800.69	15	8.20	0.01	0.00	0.01	0.01	0.02
N II	4227.74	16	7.53	0.05	0.01	0.02	0.03	0.06
	4236.99	38	7.48	0.06	0.02	0.01	0.03	0.07
	4607.16	40	7.73	0.07	0.02	0.03	0.03	0.08
	4643.09	42	7.62	0.06	0.03	0.04	0.03	0.08
	4779.72	11	7.66	0.05	0.01	0.00	0.02	0.05
	4788.14	18	7.70	0.06	0.02	0.02	0.03	0.07
	4987.38	10	7.70	0.05	0.02	0.01	0.03	0.06
	4994.36	19	7.53	0.07	0.02	0.02	0.04	0.09
	5001.30	83	7.59	0.09	0.03	0.05	0.05	0.12
	5002.70	16	7.72	0.05	0.03	0.02	0.02	0.06
	5005.17	53	7.55	0.10	0.03	0.06	0.06	0.13
	5007.33	28	7.56	0.08	0.02	0.03	0.04	0.10
	5010.62	29	7.65	0.06	0.03	0.03	0.02	0.08
	5025.66	13	7.76	0.07	0.02	0.01	0.04	0.08
	5045.10	39	7.65	0.06	0.04	0.04	0.03	0.09
	5495.65	10	7.53	0.07	0.02	0.01	0.04	0.08
	5666.63	50	7.57	0.09	0.04	0.05	0.05	0.12
	5676.01	35	7.64	0.08	0.04	0.04	0.04	0.11
	5679.55	71	7.58	0.10	0.04	0.07	0.05	0.14

continued on next page

...continued

	5686.21	21	7.50	0.07	0.03	0.02	0.03	0.08
	5710.77	33	7.76	0.08	0.04	0.04	0.04	0.11
	5747.30	8	7.59	0.06	0.03	0.01	0.03	0.07
O II	4185.44	30	8.42	0.15	0.05	0.07	0.10	0.20
	4319.78	60	8.60	0.15	0.06	0.08	0.10	0.21
	4366.70	48	8.40	0.15	0.06	0.08	0.09	0.20
	4414.90	70	8.37	0.15	0.05	0.09	0.10	0.21
	4416.97	56	8.41	0.15	0.06	0.09	0.09	0.21
	4452.38	25	8.44	0.13	0.05	0.04	0.08	0.17
	4590.97	52	8.48	0.18	0.06	0.08	0.12	0.24
	4596.07	50	8.47	0.16	0.06	0.05	0.10	0.20
	4641.81	68	8.40	0.18	0.06	0.10	0.11	0.24
	4649.13	92	8.51	0.17	0.07	0.14	0.11	0.26
	4661.63	47	8.37	0.16	0.06	0.08	0.10	0.21
	4673.73	17	8.42	0.12	0.05	0.03	0.08	0.16
	4676.23	45	8.46	0.16	0.06	0.07	0.10	0.21
	4696.32	10	8.38	0.12	0.05	0.02	0.07	0.15
	4701.44	10	8.39	0.14	0.04	0.02	0.09	0.17
	4705.35	40	8.38	0.16	0.05	0.07	0.11	0.21
	4890.85	12	8.44	0.14	0.06	0.02	0.08	0.17
	4906.83	19	8.42	0.16	0.06	0.04	0.09	0.20
	4941.07	19	8.43	0.16	0.05	0.04	0.10	0.20
	4943.00	27	8.41	0.17	0.05	0.06	0.11	0.22
	4955.71	8	8.42	0.13	0.05	0.02	0.08	0.16
	5190.49	6	8.49	0.14	0.05	0.01	0.08	0.17
	5206.64	16	8.37	0.15	0.05	0.03	0.10	0.19
Ne I	5852.49	12	8.27	0.06	0.00	0.01	0.05	0.08
	6143.06	35	8.19	0.05	0.01	0.01	0.04	0.07
	6163.59	10	8.29	0.04	0.00	0.00	0.04	0.06
	6217.28	7	8.29	0.05	0.00	0.01	0.05	0.07
	6334.43	30	8.36	0.05	0.01	0.01	0.04	0.07
	6382.99	25	8.30	0.04	0.01	0.01	0.04	0.06
	6402.25	65	8.19	0.05	0.00	0.04	0.05	0.08
	6506.53	30	8.22	0.06	0.00	0.02	0.06	0.09
Mg II	4481.20	160	7.60	0.06	0.01	0.16	0.06	0.18
Al III	4479.86	46	6.28	0.05	0.02	0.06	0.03	0.09
	4512.54	41	6.33	0.04	0.01	0.11	0.02	0.12
	4529.04	66	6.30	0.03	0.01	0.11	0.01	0.11

continued on next page

...continued

Si II	5056.15	33	7.04	0.12	0.03	0.04	0.09	0.16
	6371.37	37	7.34	0.12	0.01	0.07	0.11	0.18
Si III	4567.84	97	7.26	0.11	0.04	0.20	0.07	0.24
	4574.76	65	7.28	0.12	0.04	0.14	0.07	0.20
	4813.33	22	6.94	0.12	0.04	0.04	0.08	0.15
	4819.77	32	7.06	0.13	0.04	0.04	0.08	0.16
	4829.07	33	6.89	0.12	0.03	0.04	0.08	0.15
	5739.73	69	7.32	0.14	0.04	0.13	0.08	0.21
Fe III	4310.35	13	7.18	0.04	0.02	0.06	0.00	0.07
	4419.60	34	7.25	0.03	0.05	0.14	0.01	0.15
	5063.42	9	7.26	0.03	0.04	0.02	0.01	0.05
	5086.70	18	7.34	0.03	0.04	0.07	0.01	0.09
	5127.51	50	7.31	0.04	0.05	0.10	0.00	0.12
	5156.11	40	7.33	0.06	0.04	0.17	0.01	0.18
	5193.91	12	7.33	0.03	0.04	0.04	0.01	0.06
	5218.10	5	7.39	0.11	0.04	0.01	0.07	0.14
	5235.66	14	7.25	0.11	0.06	0.06	0.05	0.15
	5243.31	30	7.26	0.13	0.06	0.13	0.07	0.21
	5276.48	16	7.22	0.11	0.06	0.07	0.06	0.16
	5282.30	20	7.26	0.12	0.06	0.09	0.06	0.17
	5284.83	5	7.29	0.07	0.03	0.02	0.03	0.08
	5298.11	4	7.29	0.08	0.02	0.01	0.04	0.09
	5299.93	14	7.32	0.11	0.05	0.06	0.06	0.15
	5302.60	14	7.23	0.12	0.05	0.04	0.06	0.15
	5306.76	9	7.23	0.10	0.05	0.03	0.05	0.13
	5460.80	9	7.23	0.09	0.05	0.02	0.04	0.11
	5485.52	10	7.23	0.09	0.05	0.02	0.04	0.11
	5573.42	11	7.20	0.09	0.06	0.04	0.04	0.12
	5833.94	26	7.55	0.10	0.04	0.17	0.06	0.21
	5854.62	6	7.34	0.11	0.04	0.04	0.06	0.14
	6032.64	14	7.46	0.02	0.02	0.05	0.01	0.06
	6036.55	6	7.37	0.10	0.05	0.04	0.05	0.13

Comm. in Asteroseismology
Volume 162, February 2011
© Austrian Academy of Sciences

Frequency analysis of five short periodic pulsators

A. Liakos & P. Niarchos

National and Kapodistrian University of Athens, Faculty of Physics,
Department of Astrophysics, Astronomy and Mechanics,
GR 157 84, Zografos, Athens, Hellas

Abstract

CCD observations of five newly discovered pulsating stars are analyzed using the Period04 software for a comprehensive investigation of their pulsation properties. The type of their variability is discussed, and the pulsation frequencies and Fourier spectra are presented.

Accepted: February 22, 2011

Individual Objects: TYC 1134-414-1, TYC 1270-926-1, TYC 1626-1303-1, TYC 3164-1517-1, TYC 4559-2536-1

1.   Introduction, Observations and data reduction

TYC 1134-414-1 (=HIP108139 = HD 208238) was discovered as pulsating variable by Turner et al. (2008). They reported a dominant period of 0.048622 days and a spectral class of A3, therefore the star was classified as δ Scuti type pulsator.

The light variability of the stars: TYC 1270-926-1 (=GSC 01270-00926), TYC 1626-1303-1 (=GSC 01626-01303), TYC 3164-1517-1 (=BD+43 3593 = GSC 03164-01517) and TYC 4559-2536-1 (=GSC 04559-02536) was reported by Liakos & Niarchos (2011) as by-product in the field of view of other targets. TYC 3164-1517-1 is classified as F0 type star according to the AGK3 Catalogue (Heckmann 1975).

Preliminary results of the analysis of their light variations showed that low-amplitude and relatively quick periodic pulsations (45 min − 3 hrs) occur. Such a variability corresponds to typical pulsational behaviour usually met in δ Scuti and β Cephei stars, which are of A-F and B spectral type, respectively. Therefore, after the discovery of these variables, we performed follow up observations using appropriate binning modes and exposure times in order to obtain a comprehensive study of their variations. Given that these kinds of stars are hot in general, their largest pulsation amplitude is expected to occur in B-band, thus, the B filter was selected for the observations. The time resolution along with the highest possible photometric signal-to-noise ratio (S/N) was set as first priority.

The observations were performed at the Gerostathopoulion Observatory of the University of Athens from July 2009 to December 2010 using various instrumentation setups. In particular, we used a 40-cm Cassegrain telescope (T_1), a 25-cm (T_2) and a 20-cm (T_3) Newtonian reflector telescope, and the ST-10XME (C_1) and ST-8XMEI (C_2) CCDs equipped with the Bessell B, V, R, I photometric filters.

Differential magnitudes with the method of aperture photometry were obtained using the software *MuniWin* v.1.1.26 (Hroch 1998). In Table 1 we list: The number of nights (N) of observations in a given time range (T), namely the time difference in days between the last and first observation nights, the total time span ($T.S.$), the standard deviation ($S.D.$) of the observations (mean value), the abbreviation of the instrumentation (Ins) and the comparison (C) and check (K) stars used for each case. The spectral types of the stars were estimated according to their B-V indexes by using the tables published by Cox (2000), while the absolute magnitude M_v was calculated from the relation given by McNamara & Feltz (1978):

$$M_v = -3.25(12) * logP - 1.45(11) \tag{1}$$

where P is the fundamental pulsation period in days. In Table 2 the apparent magnitudes in B and V passbands as given in *The Tycho-2 Catalogue* (Hog et al. 2000), the calculated B-V index, the spectral type range and the absolute magnitude of each star are presented.

Table 1: The observations log

Star	N [nights]	T [days]	T.S. [hrs]	S.D. [mmag]	Ins	Comparison Stars
TYC 1134-414-1	5	12	22.4	5.1	$T_1 + C_1$	C: TYC 1134-247-1
						K: TYC 1134-735-1
TYC 1270-926-1	4	28	27.7	7.1	$T_2 + C_2$	C: TYC 1274-1358-1
						K: TYC 1274-1387-1
TYC 1626-1303-1	7	9	26.7	5.1	$T_1 + C_1$	C: TYC 1626-1290-1
						K: TYC 1626-1275-1
TYC 3164-1517-1	23	54	130.1	5.7	$T_1 + C_1$	C: TYC 3164-0083-1
						K: TYC 3164-0269-1
TYC 4559-2536-1	9	21	40.4	7.9	$T_3 + C_2$	C: TYC 4559-0189-1
						K: SAO 8107

Table 2: Magnitude information and spectral type estimations of the stars

Star	B	V	B-V	Sp. Type	M_v
TYC 1134-414-1	9.51 (2)	9.24 (2)	0.27 (3)	A8-F0	2.9 (5)
TYC 1270-926-1	12.3 (2)	11.5 (2)	0.8 (3)	F7-K4	2.0 (5)
TYC 1626-1303-1	12.2 (2)	12.3 (2)	-0.1 (3)	O2-A7	3.5 (5)
TYC 3164-1517-1	10.85 (5)	10.53 (6)	0.32 (8)	A8-F4	1.5 (5)
TYC 4559-2536-1	12.6 (2)	12.3 (2)	0.3 (3)	A1-G1	2.5 (5)

2. Results of frequency analyses

Frequency-analysis of all available data in the interval from 0 to 80 c/d was performed. The software *PERIOD04* v.1.2, which is based on the classical Fourier analysis (Lenz & Breger 2005), was used. The typical frequency range for δ *Scuti* stars is considered to be between 3-80 c/d (Breger 2000), while frequencies between 0-3 c/d could be caused potentially by atmospheric reasons or observational drifts. After the first frequency computation, the residuals were subsequently pre-whitened for the next one. The results of the frequency search for all cases are presented in Table 3, where the identification number of the frequency (*No*), the frequency value(*F*), its corresponding amplitude (*A*) and phase (Φ) and the signal-to-noise ratio (*S/N*) after pre-whitening for the previous frequency(ies) are listed. The error for each value, given in parenthesis, was calculated from the programme, which uses the least squares method. In addition, the sum of the squared residuals (χ^2) derived from a multi-parameter least-squares fit of sinusoidal functions, is also given for each adopted solution. The frequency spectra, the Fourier fits on the observational points for the longest (data) sets of observations and the spectral window of each star are illustrated in Figs. 1-4.

Table 3: The frequency analysis results

No	TYC 1270-926-1				TYC 1626-1303-1			
	F	A	Φ	S/N	F	A	Φ	S/N
	[c/d]	[mmag]	[deg]		[c/d]	[mmag]	[deg]	
f_1	11.437 (1)	18.1 (7)	34 (2)	19.2	32.682 (7)	3.6 (3)	279 (5)	10.1
f_2	0.366 (1)	15.8 (7)	280 (3)	47.5	39.751 (9)	2.6 (3)	223 (7)	8.7
f_3	10.643 (1)	13.0 (7)	15 (3)	16.1	33.985 (7)	3.3 (3)	355 (5)	9.2
f_4	15.334 (1)	12.1 (7)	306 (3)	14.4	32.869 (7)	3.3 (3)	78 (5)	9.7
f_5	21.050 (2)	5.9 (7)	63 (7)	6.8	37.653 (10)	2.5 (3)	157 (7)	8.1
f_6	17.063 (2)	6.9 (7)	16 (6)	9.6	68.636 (14)	1.7 (3)	54 (11)	4.4
f_7	2.339 (1)	8.0 (7)	286 (5)	9.8	24.270 (14)	1.7 (3)	218 (11)	4.1
f_8	24.548 (3)	4.1 (7)	9 (10)	7.2	42.633 (16)	1.5 (3)	280 (12)	4.4
f_9	72.964 (3)	3.4 (7)	186 (12)	4.0	–	–	–	–
f_{10}	5.653 (2)	5.2 (7)	98 (8)	6.7	–	–	–	–
χ^2	0.002				0.005			

No	TYC 1134-414-1				TYC 4559-2536-1			
	F	A	Φ	S/N	F	A	Φ	S/N
	[c/d]	[mmag]	[deg]		[c/d]	[mmag]	[deg]	
f_1	21.365 (2)	3.0 (1)	201 (3)	24.5	16.516 (1)	16.9 (5)	347 (2)	29.0
f_2	21.775 (3)	2.3 (1)	189 (3)	19.7	17.156 (1)	21.4 (5)	164 (1)	41.1
f_3	7.736 (6)	1.1 (1)	337 (7)	5.2	17.042 (1)	15.2 (5)	37 (2)	28.7
f_4	19.237 (6)	1.2 (1)	217 (7)	9.8	19.524 (2)	6.3 (5)	181 (4)	9.4
f_5	14.852 (8)	0.8 (1)	257 (9)	3.0	23.186 (3)	4.2 (5)	308 (7)	5.7
f_6	26.699 (9)	0.8 (1)	215 (10)	4.5	–	–	–	–
χ^2	0.003				0.010			

No	TYC 3164-1517-1			
	F	A	Φ	S/N
	[c/d]	[mmag]	[deg]	
f_1	8.226 (3)	8.9 (2)	120 (1)	12.8
χ^2	0.010			

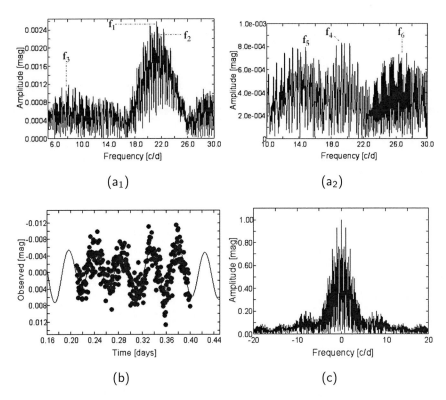

Figure 1: (a₁), (a₂): The Frequency spectra where the identified frequencies are indicated, (b): the Fourier fit on the observational points for the longest (data) set of observations, and (c): the spectral window for TYC 1134-414-1.

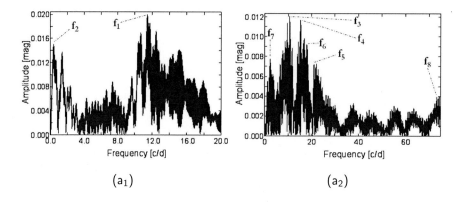

(a₁)   (a₂)

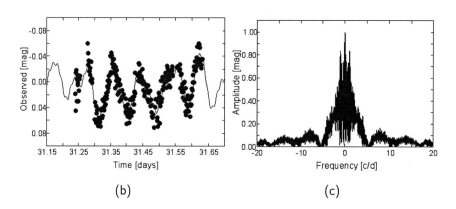

(b)   (c)

Figure 2: The same as Fig. 1 but for TYC 1270-926-1.

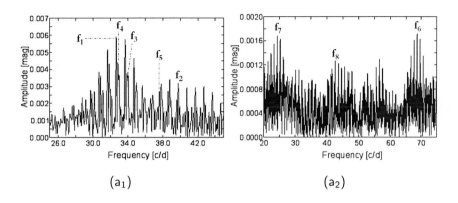

(a₁)   (a₂)

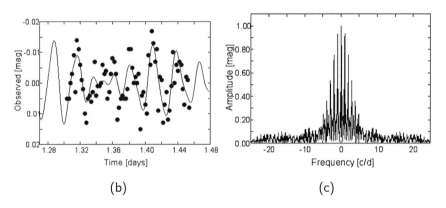

(b)   (c)

Figure 3: The same as Fig. 1 but for TYC 1626-1303-1.

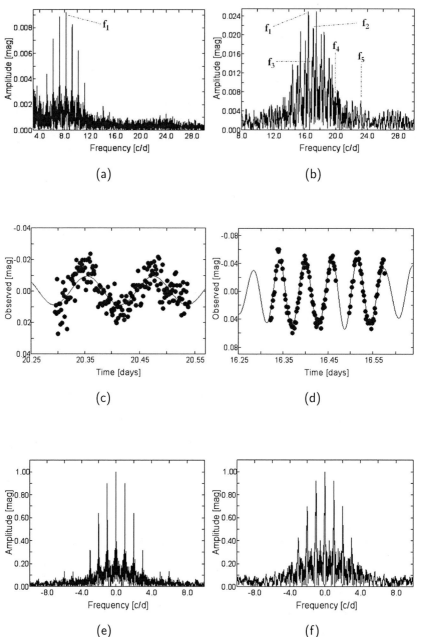

Figure 4: The same as Fig. 1, but (a), (c), (e) for TYC 3164-1517-1, and (b), (d), (f) for TYC 4559-2536-1.

TYC 1134-414-1: Six possible pulsation frequencies were obtained for this star. f_1 and f_2 appear to have very close values, constituting very probably a close frequency pair, since the criterion of Loumos & Deeming (1978) is well fulfilled. In particular, the half difference of f_1 and f_2 yields a value of $\sim$0.21 c/d, which is larger than the achieved frequency resolution of $1.5/T = 0.125$ c/d, according to our time range (see Table 1).

TYC 1270-926-1: The frequency analysis revealed ten oscillating frequencies for this star. f_2 and f_7 are not inside the typical range for δ Scuti stars, therefore their origin could be non-physical (i.e. observational drifts, atmospheric conditions). Moreover, the ratio $f_4/f_1 \sim$0.75 is typical for the radial fundamental and first overtone modes.

TYC 1626-1303-1: In this case, eight frequencies were traced. The values of f_1 and f_4 differ by only 0.187 c/d, and contrary to TYC 1134-414-1, in this case a frequency resolution of $\sim$0.17 c/d was yielded. This value questions the real separation of these frequencies, since the criterion of Loumos & Deeming (1978) is not satisfied. Due to this controversial issue, aliasing effects might implicate the results, hence another solution might be possible.

TYC 3164-1517-1: This 'slow' pulsator, unlike the other cases, was found to oscillate with a unique frequency of 8.226 c/d.

TYC 4559-2536-1: The analysis for this star resulted in five frequencies. f_2 and f_3 yield a difference of $\sim$0.1 c/d, they met the Loumos & Deeming (1978) criterion, and therefore can be considered as a close frequency pair (Breger & Bischof 2002).

3. Discussion and conclusions

In the present study five recently discovered pulsating stars are analysed with the Fourier method. Moreover, an approximate spectral classification is proposed. The combination of the above information provides the means for a coherent study of their intrinsic pulsational attributes and their classification as variable stars.

TYC 1134-414-1 was found to have six pulsation frequencies indicating a multi-periodic pulsator with radial and non-radial modes. Its spectral type, according to its B-V index, is ranging between A8-F0, but a direct spectral observation of Turner et al. (2008) resulted into an A4/5 class. Our results are in agreement with those of Turner et al. (2008) as far as the dominant period is concerned but, in addition, more pulsation frequencies are presented in the current study, due to our larger time span. The star has already been classified as δ Scuti type.

TYC 1270-926-1 is a slight ambiguous case for classification. Although its 8-mode pulsational behaviour is certain, its spectral type was found to range between F7-K4. This value indicates a rather cool star, therefore it cannot be classified as a classical early-type pulsator.

As in the first case, TYC 1626-1303-1 is a multi-oscillating star, with eight frequencies identified. Its spectral type was found to be O2-A7, which partially covers both the range of δ Scuti and β Cephei pulsating stars, but its absolute magnitude leads us to the conclusion that it could be a δ Scuti type pulsator (see Fig. 5).

TYC 3164-1517-1 has a spectral type between A8-F4 according to its B-V index. This result is consistent with its previously F0 classification (Heckmann 1975), placing it well inside the δ Scuti range. The frequency analysis revealed its mono-periodic oscillation.

TYC 4559-2536-1 can be classified as a δ Scuti pulsator with five frequencies in radial and non-radial modes, and a spectral type between A1-G1.

In Fig. 5 the position of each star with its uncertainties in the Colour-Magnitude diagram is illustrated. The large errors come from the large uncertainty in their spectral types, derived from the B-V values.

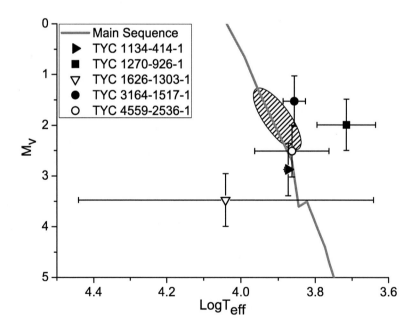

Figure 5: The location of the stars (symbols) in the Colour-Magnitude diagram. The dark grey solid line represents the Main Sequence, while the shaded ellipse corresponds to the δ Scuti region.

For all stars, except TYC 1134-414-1 and TYC 3164-1517-1, spectroscopic observations will certainly improve the present classification of the stars, while photometric observations using larger telescopes in a larger time span will probably reveal more frequencies, which are below the detection threshold of the current instrumentation used for each case. In addition, for the case of TYC 1626-1303-1, where an ambiguous close pair of frequencies was found, photometric observations covering a larger time range will prove or disprove their separation.

Acknowledgments. This work has been financially supported by the Special Account for Research Grants No 70/4/9709 of the National & Kapodistrian University of Athens, Hellas. The authors would like to thank the anonymous referee for valuable comments that improved significantly the present work.

References

Breger, M. 2000, ASPC, 210, 3

Breger, M. & Bischof, K.M. 2002, A&A, 385, 537

Cox, A.N. 2000, *Allen's Astrophysical Quantities 4th ed.*, Springer-AIP press

Heckmann O. 1975, *AGK 3. Star catalogue of positions and proper motions north of -2.5 deg. declination*, Hamburg-Bergedorf: Hamburger Sternwarte, 1975, edited by Dieckvoss, W.

Hog, E., Fabricius, C., Makarov, V.V., et al. 2000, A&A, 355, 27

Hroch, F. 1998, Proceedings of the 29th Conference on Variable Star Research, 30

Liakos, A. & Niarchos, P. 2011, IBVS, 6000, No 1, 2, 6 and 10, in press

Lenz, P. & Breger, M. 2005, CoAst, 146, 53

Loumos, G.L. & Deeming, T.J. 1978, ApSS, 56, 285

McNamara, D.H. & Feltz, K.A., Jr. 1978, PASP, 90, 275

Turner, G., Kaitchuck, R., & Skiff, B. 2008, JAVSO, 36, 156

Comm. in Asteroseismology
Volume 162, June 2011
© Austrian Academy of Sciences

Update and Additional Frequencies for *Kepler* Star KIC 9700322

J. A. Guzik[1] & M. Breger[2,3]

[1] Los Alamos National Laboratory, XTD-2, MS T086, Los Alamos, NM 87545-2345, USA

[2] Institut für Astronomie der Universität Wien, Türkenschanzstr. 17, A–1180, Wien, Austria

[3] Department of Astronomy, University of Texas, Austin, TX 78712, USA

Abstract

Breger et al. (2011) reported on the discovery of 76 frequencies for the star KIC 9700322 from one month (Sept.-Oct. 2009) of short-cadence (1-minute sampling rate) photometric data during Quarter 3 of the NASA *Kepler* mission. Here we report on a reanalysis combining an additional month (June-July 2009) obtained during Quarter 2 as part of the *Kepler* Guest Observer program. This analysis confirms all but two of the earlier frequencies > 0.5 c/d, and in addition finds six new combination frequencies of the 8 highest amplitude modes, and three frequencies that are not combinations of these modes. Since we do not know the parents of 12 of the 83 frequencies, additional astrophysical discoveries may await in continued study of this star. The rotational modulations of the two radial modes and the $\ell = 2$ quintuplet are confirmed. For the two radial modes (f_1, f_2) and the central $(m = 0)$ mode of the $\ell = 2$ quintuplet, f_6, amplitude variability and/or close frequencies were found. The additional data do not reveal more of the $\ell=0$, 1, 2 or higher-degree p modes that were expected, and this result is still a mystery.

Accepted: June 27, 2011

Individual Objects: KIC 9700322

1. Introduction

As part of its search for Earth-sized planets around sun-like stars, the *Kepler* mission is surveying around 150,000 stars for planetary transits. In the process, it is acquiring photometry with unprecedented precision and continuous length of time series on hundreds of pulsating variable stars, including many δ Scuti stars (Borucki et al. 2010; Gilliland et al. 2010).

The δ Scuti star KIC 9700322 is a slowly rotating, cool δ Scuti star ($T_{eff} = 6700 \pm 100$ K, $\log g = 3.7 \pm 0.1$, $v \sin i = 19 \pm 1$ km/s; Breger et al. 2011). δ Sct stars are variable stars of A-F spectral type on or near the main sequence that show radial and nonradial oscillations with frequencies $\sim$5-30 cycles/day. The pulsations are driven by the "κ effect", a feedback mechanism produced by the opacity increase in the region of second helium ionization in the stellar envelope (see, e.g., Aerts et al. 2010). The brightness variations of KIC 9700322 were measured by the *Kepler* satellite for 30.3 d during Quarter 3. Analyses of these data revealed 76 frequencies with amplitudes as small as 14 ppm. Two dominant radial modes, an $\ell = 2$ quintuplet and a brightness modulation during rotation were detected (i.e., eight frequencies). Almost all other detected frequencies could be identified with various combinations and rotational modulations of these modes. Good agreement of the frequency spacings with theoretical models could be obtained (Breger et al. 2011, hereafter called Paper I). A remarkable result was the absence of additional independent frequencies down to an amplitude limit near 14 ppm, suggesting that the star is stable against most forms of pulsation.

In 2008, as part of the *Kepler* Guest Observer Cycle 1 program, before *Kepler* was launched, J.A. Guzik proposed to observe KIC 9700322 along with thirteen other stars in a search for hybrid γ Doradus/δ Sct stars, and the data was taken in Q2. Because KIC 9700322 did not show hybrid behavior with both higher (10-15 c/d) and low (1-5 c/d) frequencies, this star was not considered further until the Paper I preprint appeared. Since we realized that we had additional short cadence data readily available, we decided to combine the two quarters of data to see whether the results from one quarter could be confirmed and extended.

2. The Q2 *Kepler* spacecraft observations of KIC 9700322

Portions of the Q2 "raw" data (that have been processed to reduce the effects of bias and dark current, and flat-fielded) have small systematic zero-point drifts in brightness. These shifts are removed in the "corrected" data provided by the *Kepler* team, but these data that have been pre-conditioned to optimize the search for planetary transits (see http://keplergo.arc.nasa.gov/PipelinePDC.shtml) are not suitable for asteroseismological investigations, and so were not used here. In the Q2 data for KIC9700322, a two-day stretch immediately after an observing gap near HJD 245 5016 contains a large drift reaching several parts per thousand. These

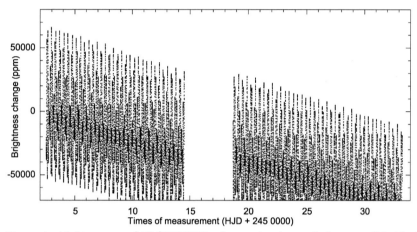

Figure 1: Light curve of KIC9700322 after reductions of the new Q2 *Kepler* data.

data, from HJD 245 5016.71 to 245 5018.71, were not used. For the remaining data, the small remaining zero-point drifts were removed artificially, similar to the techniques applied by Garcia et al. (2011). This processing affects only the analysis of the low-frequency region. Consequently, the present paper covers only periods shorter than two days, or $f > 0.5$ c/d, i.e., the period range of the typical δ Sct stars. As a final step, the data were filtered by us for obvious outliers (deviations larger than 4 σ). Fig. 1 shows the new observations after our reductions.

3.   Frequency analysis

The KIC 970032 *Kepler* data were analyzed with the statistical package PERIOD04 (Lenz & Breger 2005). This package carries out multifrequency analyses with Fourier as well as least-squares algorithms. White noise is not assumed. Following the standard procedures for examining the peaks with PERIOD04, we have determined the amplitude signal/noise values for every promising peak in the amplitude spectrum and adopted a limit of S/N of 4.0. As in the previous paper on this star, the noise is calculated from prewhitened data because of the huge range in amplitudes of three orders of magnitude.

It was found that the Q2 data essentially confirms (and extends) the results previously obtained from the Q3 data alone. The amplitude spectra of Q2 and Q3 are essentially identical to each other (except for alias structure in Q2 because of the gap in the middle of Q2). This demonstrates the stability of the pulsation of KIC 9700322. The matching of the Q2 and Q3 data also shows that the quality of the data in the two observing quarters is comparable, despite the more significant instrumental systematics in the Q2 data. Due to the similarity, new figures of the amplitude spectra are not shown in this paper and we refer to the figures in Paper I. We will now concentrate on the analysis of the combined Q2 and Q3 data, which covers 121d with a central gap of 60d. The long time base improves the frequency resolution and allows us to look for amplitude variability and close frequencies. Also, the additional data allow us to detect several additional frequencies.

79 statistically significant frequency peaks were detected in the frequency range > 0.5 cycle c/d. These frequencies are listed in Table 1, along with four frequencies < 0.5 c/d that were reported using the Q3 data in Paper 1. Since the rotational frequency, $f_3 = 0.1597$ c/d, is important for modulations of the amplitudes of most (all?) pulsation modes, the rotational frequency and its multiples are also listed in the table with the values taken from Paper I.

Of the 9 new frequencies detected in the combined Q2Q3 analysis, 6 are easily identified as combination modes of the main frequencies. One frequency forms a close double with f_6 (see below). On the other hand, the small-amplitude peak from Paper I at 51.752 c/d is no longer statistically significant, while the unconfirmed peak at 12.584 c/d probably arose from the slow amplitude modulation of f_2 (see below).

71 of the 83 frequencies listed in Table 1 can be identified with f_1 - f_8 and their combinations. Only 12 frequencies cannot be identified in this manner. However, these 12 frequencies are not all additional, independent pulsation modes. It is interesting that most of these frequencies are numerically related to each other through new combinations, e.g., the difference between the two peaks at 12.5333 and 22.3258 c/d is exactly f_1. Since 12.5333 c/d is a close double to f_2, 22.3258 c/d is the identical close double to the $(f_1 + f_2)$ peak.

Table 1: Multifrequency solution of KIC 9700322 and identifications.

Frequency c/d	μHz	Identification	Amplitudes in ppm Q2Q3	Q2	Q3	Comment
Main frequencies						
9.7925	113.339	f_1	27302	27343	27266	Radial
12.5688	145.472	f_2	29449	29463	29440	Radial
0.1597	1.848	f_3			80	Rotation
11.3187	131.003	f_4	27	25	29	Quintuplet
11.4551	132.582	f_5	147	147	147	Quintuplet

Continued...

...continued from previous page

Frequency		ID	Amplitudes in ppm			Comment
c/d	μHz		Q2Q3	Q2	Q3	
11.5906	134.150	f_6	465	474	460	Quintuplet
11.7201	135.649	f_7	215	207	222	Quintuplet
11.8576	137.241	f_8	115	118	112	Quintuplet
Combination frequencies						
0.3194	3.697	$2f_3$			174	2x rotation
0.4791	5.545	$3f_3$			33	3x rotation
0.6388	7.394	$4f_3$	17	13	25	
0.7112	8.232	$f_2 - f_8$	16	19	26	
0.9782	11.322	$f_2 - f_6$	31	36	26	
1.1137	12.890	$f_2 - f_5$	48	55	46	
1.6625	19.242	$f_5 - f_1$	20	16	25	
1.7980	20.811	$f_6 - f_1$	27	39	17	
2.7763	32.133	$f_2 - f_1$	2640	2647	2633	
2.9360	33.981	$f_2 - f_1 + f_3$	22	18	26	
5.5526	64.266	$2f_2 - 2f_1$	86	84	89	
7.0162	81.206	$2f_1 - f_2$	630	628	632	
9.4731	109.643	$f_1 - 2f_3$	50	47	58	
9.6328	111.491	$f_1 - f_3$	52	54	51	
9.9522	115.188	$f_1 + f_3$	57	63	53	
10.1119	117.036	$f_1 + 2f_3$	39	37	38	
12.2494	141.776	$f_2 - 2f_3$	34	35	35	
12.4091	143.624	$f_2 - f_3$	36	37	32	
12.7285	147.321	$f_2 + f_3$	33	34	32	
12.8882	149.169	$f_2 + 2f_3$	21	18	23	
15.3451	177.605	$2f_2 - f_1$	497	492	503	
16.8088	194.546	$3f_1 - f_2$	67	64	70	
18.1214	209.738	$3f_2 - 2f_1$	14	16	11	
19.5850	226.679	$2f_1$	2226	2227	2225	
21.2476	245.921	$f_5 + f_1$	45	48	42	
21.3831	247.489	$f_6 + f_1$	40	46	36	
21.5126	248.988	$f_7 + f_1$	18	22	16	
21.6501	250.580	$f_8 + f_1$	18	17	19	
22.2016	256.963	$f_1 + f_2 - f_3$	26	26	26	
22.3613	258.812	$f_1 + f_2$	4900	4899	4901	
22.5210	260.660	$f_1 + f_2 + f_3$	21	25	18	
22.7738	263.585	$f_4 + f_5$	13	11	15	

Continued...

...continued from previous page

Frequency		ID	Amplitudes in ppm			Comment
c/d	μHz		Q2Q3	Q2	Q3	
23.0456	266.732	$f_5 + f_6$	34	31	39	
23.1811	268.300	$2f_6$	10	15	6	
23.7152	274.481	$2f_8$	25	21	28	
24.0239	278.054	$f_5 + f_2$	34	34	34	
24.1594	279.622	$f_6 + f_2$	47	48	46	
24.2889	281.121	$f_7 + f_2$	52	51	52	
24.4264	282.713	$f_8 + f_2$	54	52	56	
24.9779	289.096	$2f_2 - f_3$	17	21	16	
25.1376	290.945	$2f_2$	2661	2658	2663	
26.6013	307.885	$4f_1 - f_2$	13	14	12	
27.9139	323.078	$3f_2 - f_1$	203	203	203	
29.3776	340.018	$3f_1$	190	188	191	
32.1538	372.151	$2f_1 + f_2$	482	485	479	
33.8164	391.393	$f_1 + f_2 + f_5$	12	15	10	
33.9519	392.962	$f_1 + f_2 + f_6$	15	14	16	
34.2189	396.052	$f_1 + f_2 + f_8$	17	19	16	
34.9301	404.284	$f_1 + 2f_2$	536	536	536	
37.7064	436.417	$3f_2$	329	329	329	
39.1701	453.357	$4f_1$	19	23	16	
40.4827	468.550	$4f_2 - f_1$	34	34	34	
41.9464	485.490	$3f_1 + f_2$	22	22	23	
44.7227	517.623	$2f_1 + 2f_2$	113	112	114	
47.4989	549.756	$f_1 + 3f_2$	78	74	82	
50.2752	581.889	$4f_2$	82	80	83	
51.7389	598.830	$4f_1 + f_2$	9	8	10	
54.5152	630.963	$3f_1 + 2f_2$	35	34	35	
57.2915	663.096	$2f_1 + 3f_2$	55	53	58	
60.0677	695.229	$f_1 + 4f_2$	34	34	34	
62.8440	727.361	$5f_2$	18	21	15	
67.0840	776.435	$3f_1 + 3f_2$	19	18	19	
69.8603	808.568	$2f_1 + 4f_2$	19	16	21	
Peaks detected in Q3 with frequency < 0.5 c/d						
0.0221	0.256	f_{72}			347	
0.0555	0.642	f_{73}			95	
0.1346	1.558	f_{74}			35	
0.3542	4.100	f_{75}			25	

Continued...

...continued from previous page

Frequency		ID	Amplitudes in ppm			Comment
c/d	μHz		Q2Q3	Q2	Q3	
Other peaks in the amplitude spectrum						
8.1394	94.206	f_{76}	15	11	18	
11.5806	134.035	f_{77}	135	140	132	
11.9712	138.555	f_{78}	16	19	13	
12.5333	145.061	f_{79}	61	37	80	
13.2406	153.247	f_{80}	18	16	21	
14.6253	169.274	f_{81}	14	10	17	
22.3258	258.400	f_{82}	15	8	20	
24.1493	279.506	f_{83}	45	48	42	

4.   Amplitude variability/close frequencies

Amplitude variability and the presence of close-frequency pairs (which lead to amplitude variability in short data sets) were examined in two ways. To examine the amplitude variability directly, we have subdivided the data into three-day groups. For each group we have computed the optimum amplitudes for the dominant modes, while using a common solution for the other frequencies. The second method relied on Fourier analyses and multiple-least-squares solutions in PERIOD04 to look for close frequencies.

For the two radial modes (f_1, f_2) and the central ($m = 0$) mode of the $\ell = 2$ quintuplet, f_6, amplitude variability and/or close frequencies were found, that we discuss below.

4.1. $f_6 = 11.591$ c/d

In the previous paper we identified f_6 as the axisymmetric mode ($m = 0$) of an ℓ=2 quintuplet. When we add the Q2 data, it becomes obvious that the mode has a variable amplitude, possibly caused by a close companion frequency. The amplitude variability is shown in Fig. 2. If we assume a sinusoidal amplitude variation, we obtain a modulation frequency of 0.010 ± 0.001 c/d or a modulation period near 100d.

This 100d modulation should also be visible in the Fourier spectrum. In the individual Q2 or Q3 data sets alone, the frequency resolution is not sufficient to reveal the double structure. When the two data sets are combined, the double frequency structure becomes apparent. Some uncertainties due to aliasing (structures separated by 0.011 c/d) still remain. When we prewhiten

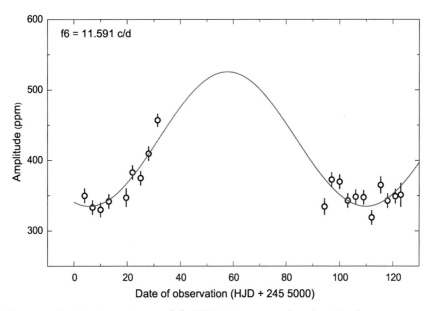

Figure 2: Amplitude variation of f_6. This is interpreted as beating between two close frequencies.

the dominant frequency of 11.591 c/d, a number of peaks remain in the amplitude spectrum. The peak at 11.581 c/d for the second peak leads to a clean power spectrum after prewhitening. An alias peak at 11.569 c/d, on the other hand, is unsatisfactory because it leaves a number additional close peaks and leads to higher residuals. We, therefore, adopt 11.581 c/d for the second peak. Our adopted frequency difference between the two frequencies in the doublet is in agreement with the value derived above from inspecting the amplitude variation.

Furthermore, the fact that we obtain only two statistically significant peaks (as opposed to a triplet with the main frequency in the middle) in the power spectrum supports the interpretation as beating between two close frequencies, rather than plain amplitude variability. Nevertheless, we must caution against potential overinterpretation of our data, since the length of data set is similar to the detected modulation period.

4.2.   Radial mode f_1

We have examined potential amplitude variation in the same manner as for f_6. The radial mode f_1 varies by about 0.2% of its value with a period of about 125d ($f_{mod} = 0.0081 \pm 0.004$ c/d). Since the modulation amplitude is very small and the modulation period is similar to the length of the data set, Fourier analyses of this modulation remain inconclusive.

4.3. Radial mode f_2

In the Q3 data alone we detected a second frequency, separated from f_2 by about 0.03 c/d. We confirm the existence of this frequency, for which an improved value of 12.533 c/d is determined. The amplitudes of f_2 suggest that an additional small amplitude modulation of f_2 with a time scale in excess of 150d also exists. Due to the small size of this variation we cannot obtain a more quantitative result at this stage.

5. Discussion

KIC 9700322 is remarkably stable compared to some δ Sct stars that show larger frequency amplitude variations over periods of weeks (see e.g., Breger 2009 and discussion on the δ Sct star 4 CVn.) KIC 9700322 shares with other stars with amplitude variability a long-term modulation of hundreds of days. Understanding the reasons for these amplitude and frequency variations requires long-term monitoring and high photometric precision. There are other mysteries associated with KIC 9700322 that are also present for other δ Sct stars: Why, for example, does this star show only a couple of radial modes and one $\ell=2$ p-mode, and not other expected low-degree modes? Which (invisible?) frequencies are the parents of the new pulsation modes that are not combinations of the modes of highest amplitude? Why does this star not show any lower-frequency γ Dor pulsations as do most of the A and early F main sequence stars observed by *Kepler* (Grigahcéne et al. 2010; Balona et al. 2011)? KIC 9700322 continues to be monitored, and analysis of an additional 3 months of short cadence data from Q6 will begin soon. In addition, long-cadence (29.4-minute integration time) data exist for Q0 through Q7, so future work may be able to explore longer-term trends. Hopefully, an analysis of KIC 9700322 in comparison and contrast with the many other γ Dor and δ Sct stars being observed by *Kepler* will help to solve some of the puzzles associated with the pulsation behavior of these stars.

Acknowledgments. This investigation has been supported by the Austrian Fonds zur Förderung der wissenschaftlichen Forschung through project P 21830-N16. We thank the entire *Kepler* team for making this data possible. J.G. is pleased to acknowledge support and data acquisition from the *Kepler* Guest Observer Cycle 1 program grant number KEPLER08-0013.

References

Aerts, C., Christensen-Dalsgaard, J. & Kurtz, D. W. 2011, *Asteroseismology*, Astronomy and Astrophysics Library, Springer, pp. 49-56

Balona, L., Guzik, J. A., Uytterhoeven, K., et al. 2011, "The *Kepler* View of γ Dor Stars", MNRAS, in press

Borucki, W. J. , Koch, D., Basri, G., et al. 2010, Science, 327, 977

Breger, M., Balona, L., Lenz, P., et al. 2011, MNRAS, 414
doi: 10.1111/j.1365-2966.2011.18508 (Paper I)

Breger, M. 2009, "Period Variations in δ Scuti Stars," in Stellar Pulsation: Challenges for Theory and Observation, eds. J.A. Guzik and P.A. Bradley, AIPC conference proceedings Volume 1170, pp. 410-414

Garcia, R. A., Hekker, S., Stello, D. et al. 2011, MNRAS, 414, L6

Gilliland, R. L., Brown, T. M., Christensen-Dalsgaard, J., et al. 2010, PASP, 122, 131

Grigahcéne, A., Antoci, V., Balona, L., et al. 2010, ApJ, 713, 192

Lenz, P. & Breger, M. 2005, CoAst, 146, 53

Comm. in Asteroseismology
Volume 162, July 2011
© Austrian Academy of Sciences

New pulsation analysis of the oscillating Eclipsing Binary BG Peg

A. Liakos & P. Niarchos

National and Kapodistrian University of Athens, Faculty of Physics,
Department of Astrophysics, Astronomy and Mechanics,
GR 157 84, Zografos, Athens, Hellas

Abstract

New BVRI light curves of the eclipsing binary BG Peg were obtained and analysed using the Wilson-Devinney code. The observations (102 hours during a span of 108 days) were collected to perform an accurate frequency analysis of the pulsations of the primary component of the system, which is of δ Sct type. Three pulsation frequencies were identified in that star.

Accepted: July 22, 2011

Individual Objects: BG Peg

1. Introduction

The story of the discovery of the variability of BG Peg (V=10.5 mag, P$\sim$ 1.95 days) is not very clear and is a hidden chapter in the history of variable star research. Prager & Shapley (1941) listed this system as a variable and reported that the discovery was made by Zessewitsch (1931) in Moscow, but the exact reference is missing from the literature. Budding (1984) found the mass ratio of the binary system to be 0.51, Brancewicz & Dworak (1980) and Svechnikov & Kuznetsova (1990) calculated the absolute parameters of the system's components and found different mass values. However, in these three works the system is classified as A2 type. Independently, and almost simultaneously, Soydugan et al. (2009) and Dvorak (2009) announced the pulsating nature of the primary component of the system and they found a dominant pulsating frequency of $\sim$25.5 c/d. Most recently, a comprehensive study of this eclipsing binary (EB) was made by Soydugan et al. (2011), with frequency analyses based on B and V photometry. In the same work, an orbital period study was also performed, resulting in evidence for a tertiary companion, mass loss from the system and mass transfer between the components.

In the study presented here, we present and analyse four-colour light curves covering a three-month time span, to identify pulsation frequencies in the data. Our analysis assumes a third light contribution to the system's apparent brightness.

2.   Observations and data reduction

The BVRI photometric observations were carried out during 15 nights from 6 August to 21 November 2010 at Gerostathopoulion Observatory of the University of Athens. Two instrumentation set-ups (a 40-cm Cassegrain and a 25-cm Newtonian Reflector telescopes equipped with the ST–10 XME and the ST–8 XMEI CCD cameras, respectively) were used to collect the data. Aperture photometry techniques were applied to CCD images and differential magnitudes were obtained with the software package *MuniWin* v.1.1.26 (Hroch 1998). TYC 1698-1148-1 (V=10.4 mag) and TYC 1698-1078-1 (V=11.7 mag) were used as comparison and check stars, respectively. The mean standard deviation of the data points were 5.8, 6.1, 7.3 and 8.6 mmag for B, V, R and I filters, respectively.

3.   Light curve analysis

The light curves of the BG Peg system were analysed simultaneously using the *PHOEBE* v.0.29d software (Prša & Zwitter 2005) which uses the method of the 2003 version of the Wilson-Devinney (WD) code (Wilson & Devinney 1971; Wilson 1979, 1990). The 'Multiple Subsets' procedure was used to approach our final solution. Lacking a spectroscopic mass ratio, we applied the 'q-search' method with a step of 0.1, with trials in Mode 2 (detached system) and Mode 5 (semidetached system with the secondary component filling its Roche Lobe), to find plausible 'photometric' estimates for the mass ratio q_{ph}. Based on the secondary component's potential, which fits better to a star which fills its respective Roche Lobe (cf. İbanoğlu 2006), Mode 5 was selected as the better way to describe the geometry of this system, and we find $q_{ph} \sim 0.2$. In the subsequent analysis, this value was set as initial input and treated as free parameter. The temperature of the primary was assigned a value of 9000 K according to its spectral type by using the correlation given in the tables of Cox (2000) and was fixed during the analysis. The temperature of the secondary T_2 was adjusted. The albedos, A_1 and A_2, and gravity darkening coefficients, g_1 and g_2, were set to generally adopted values (von Zeipel 1924; Lucy 1967; Rucinski 1969) for the given spectral types of the components. The linear limb

Table 1: Results of light curves solution with the corresponding errors given in parentheses.

Parameter	System related		filter depended			
i [deg]	83.7 (1)		B	V	R	I
q $(=m_2/m_1)$	0.219 (1)	L_1/L_T	0.955 (1)	0.927 (2)	0.903 (2)	0.873 (3)
	Component related	L_1/L_T	0.045 (1)	0.073 (1)	0.097 (1)	0.127 (1)
	P	S	x_1	0.536	0.460	0.380	0.300
T [K]	9000[a]	5095 (12)	x_2	0.849	0.706	0.609	0.512
Ω	3.67 (1)	2.28		fractional radii		
g[a]	1	0.32		pole	point	side	back
A[a]	1	0.5	r_1	0.289	0.297	0.293	0.295
Σres^2	0.638		r_2	0.240	0.500	0.249	0.281

[a]assumed, P=primary, S=secondary, $L_T=L_1+L_2$

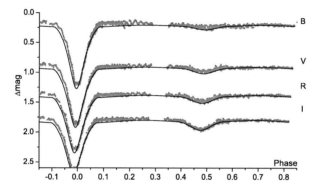

Figure 1: Synthetic (solid lines) and observed (grey points) light curves of BG Peg.

darkening coefficients, x_1 and x_2, were taken from the tables of van Hamme (1993); the dimensionless potentials Ω_1 and Ω_2, the fractional luminosity of the primary component L_1 and the inclination i of the system's orbit were set in the programme as adjustable parameters. Due to a possible existence of a tertiary component orbiting the EB (Soydugan et al. 2011), the third light option l_3 was applied, but it resulted in unrealistic values. As a result, it is not included in the final solution. Synthetic and observed light curves are presented in Fig. 1, with the corresponding parameters of the model fits given in Table 1.

4. Frequency analysis

A frequency search of the light curve residuals, except near the primary eclipse, was performed in the frequency range $3 - 80$ c/d, which is a typical range for

Table 2: The frequency analysis results of BG Peg.

	F [c/d]	A [mmag]	Φ [deg]	S/N	F [c/d]	A [mmag]	Φ [deg]	S/N
		B - filter				V - filter		
f_1	25.544 (1)	13.1 (6)	8 (3)	14.8	25.544 (1)	11.7 (6)	10 (3)	18.7
f_2	23.085 (1)	8.4 (6)	117 (4)	9.9	23.085 (1)	6.3 (6)	118 (5)	10.4
f_3	22.061 (1)	6.1 (6)	19 (6)	6.5	22.061 (1)	6.1 (6)	8 (5)	9.0
Σres^2		0.017				0.014		
		R - filter				I - filter		
f_1	25.544 (1)	8.3 (6)	359 (4)	9.8	25.543 (1)	7.1 (6)	16 (5)	9.8
f_2	23.094 (1)	4.3 (6)	147 (8)	4.7	23.085 (1)	4.0 (6)	111 (8)	5.4
f_3	22.062 (1)	3.9 (6)	9 (9)	4.1	22.060 (1)	3.3 (6)	19 (10)	4.7
Σres^2		0.015				0.015		

δ Scuti stars (Breger 2000; Soydugan et al. 2006). We used the software package *PERIOD04* v.1.2 (Lenz & Breger 2005), which is based on the classical Fourier analysis. After each frequency identification, the residuals were subsequently pre-whitened for the next one. The frequency search stopped when the detected frequency(ies) had a signal-to-noise ratio (S/N) less than 4 and an amplitude value lower than the significance limit (4σ level). An additional frequency search was made in the range $0 - 3$ c/d, for frequencies which potentially could be caused by terrestrial atmospheric effects, instrumental drifts or imperfect light curve fits. Low frequencies can also result from g-mode pulsations (Breger 2005), which are expected to be traced in the data of more than one filter. In this data set, frequencies with values less than 3 c/d were indeed detected, but they were found to be different for each filter. Thus, they cannot be considered as physically intrinsic to the BG Peg system and they are excluded from the final solution shown in Table 2.

The amplitude spectrum, the spectral window and the Fourier fit on the longest data sets are plotted in Fig. 2.

5.   Discussion and conclusions

We have reduced and analysed time series BVRI photometry of BG Peg. BG Peg was found to be a semidetached system with its cooler and less massive component filling its Roche Lobe. Therefore, the eclipsing pair, according to the definition suggested by Mkrtichian et al. (2004), is a classical oscillating Algol (oEA). A third light contribution was included in the analysis, due to a suggested existence of a tertiary component (Soydugan et al. 2011), but the results did not support it. Our light curves solution, however, is in very good agreement with that of Soydugan et al. (2011).

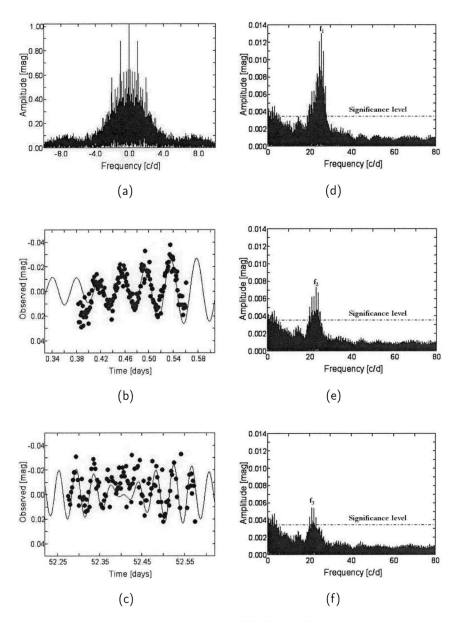

Figure 2: (a): The spectral window plot, (b)–(c) the Fourier fit on the longest data sets, and (d)–(f) the Fourier Amplitude spectrum where the detected frequencies and the significance level are indicated.

Three significant pulsation frequencies were detected in all bandpasses with decreasing amplitudes from B to I. Soydugan et al. (2011) found only two frequencies for the hotter component of BG Peg. We confirm the fundamental pulsation frequency f_1 reported by Soydugan et al. (2011), but we cannot confirm the second frequency in their study ($f_2 \sim 21.06$ c/d). The two new frequencies we have found might be because of our better time coverage: $\sim$ 102 hrs on 15 nights during 108 days for our study, versus $\sim$ 56 hrs on 13 nights during 87 days for the Soydugan et al. (2011) study.

Radial velocity measurements are certainly needed to derive the spectroscopic mass ratio and for a more coherent study of the primary component's pulsational characteristics.

Acknowledgments. This work has been financially supported by the Special Account for Research Grants No 70/4/9709 of the National & Kapodistrian University of Athens, Hellas. We used the SIMBAD database, operated at CDS, Strasbourg, France, and the Astrophysics Data System Bibliographic Services (NASA). We thank the anonymous referee for his valuable comments, which improved the quality of the paper.

References

Brancewicz, H.K. & Dworak, T.Z. 1980, AcA, 30, 501

Breger, M. 2000, ASPC, 210, 3

Breger, M. 2005, ASPCS, 333, 138

Budding, E. 1984, Bull. D'Inf. Cent. Donnees Stellaires, 27, 91

Cox, A.N. 2000, *Allen's Astrophysical Quantities 4th ed.*, Springer-AIP press

Dvorak, S. 2009, CoAst, 160, 64

Hroch, F. 1998, Proceedings of the 29th Conference on Variable Star Research, 30

İbanoğlu, C., Soydugan, F., Soydugan, E., Dervişoğlu, A., 2006, MNRAS, 373, 435

Lenz, P. & Breger, M. 2005, CoAst, 146, 53

Lucy, L.B. 1967, Zeitschrift für Astrophysik, 65, 89

Mkrtichian, D.E., Kusakin, A.V., Rodriguez, E., et al. 2004, A&A, 419, 1015

Prager, R. & Shapley, H. 1941, AnHar, 111, 1

Prša, A. & Zwitter, T. 2005, ApJ, 628, 426

Rucinski, S.M. 1969, AcA, 19, 245

Soydugan, E., Soydugan, F., Demircan, O., İbanoğlu, C. 2006, MNRAS, 370, 2013

Soydugan, E., Soydugan, F., Şenyüz, T., et al. 2009, IBVS, No 5902

Soydugan, E., Soydugan, F., Şenyüz, T., Püsküllü, Ç., Demircan, O. 2011, NewA, 16, 72

Svechnikov, M.A. & Kuznetsova, E.F. 1990, *Catalogue of Approximate Photometric and Absolute Elements of Eclipsing Variable Stars*, A.M. Gorky University of the Urals, Sverdlovsk

van Hamme, W. 1993, AJ, 106, 2096

von Zeipel, H. 1924, MNRAS, 84, 665

Wilson, R.E. & Devinney, E.J. 1971, ApJ, 166, 605

Wilson, R.E. 1979, ApJ, 234, 1054

Wilson, R.E. 1990, ApJ, 356, 613